WITHDRAWN

D1213159

PERILOUS PLACE, POWERFUL STORMS

■ ■ ■ ■

Craig E. Colten

PERILOUS PLACE, POWERFUL STORMS

Hurricane Protection in Coastal Louisiana

▪ ▪ ▪ ▪

University Press of Mississippi / Jackson

www.upress.state.ms.us

Designed by Todd Lape

The University Press of Mississippi is a member
of the Association of American University Presses.

Photograph on page ii: Flooding in the Ninth Ward,
September 1965. Courtesy National Oceanographic
and Atmospheric Administration.

First printing 2009
∞
Library of Congress Cataloging-in-Publication Data

Colten, Craig E.
Perilous place, powerful storms : hurricane protection
in coastal Louisiana / Craig E. Colten.
p. cm.
Includes bibliographical references and index.
ISBN 978-1-60473-238-2 (cloth : alk. paper)
1. Hurricane protection—Louisiana—Evaluation.
2. Flood control—Louisiana—New Orleans—Evaluation.
3. Levees—Louisiana—New Orleans—History. 4. Levees—
Mississippi River. 5. Hurricane Katrina, 2005. I. Title.

TC425.M63L73 2009
363.34'9220976335—dc22 2008047085

British Library Cataloging-in-Publication Data available

CONTENTS

ILLUSTRATIONS

FIGURES

TABLES

ACKNOWLEDGMENTS

Numerous organizations and individuals contributed to making this project possible. First and foremost, the original version of this work was done under contract with the U.S. Army Corps of Engineers History Office. Martin Reuss served as the initial project manager and capably steered this project through its beginnings. Matt Pearcy, who assumed Marty's position in spring 2006, has proved helpful as the Corps evaluated the initial draft.

Staff at the New Orleans District office also played a vital role. Despite terrible disruption to working and living conditions in the year after Hurricane Katrina, they stuck with their jobs and provided me with assistance. Gary Hawkins, in particular, assisted my field researchers in gaining access to documents. Ed Lyon offered insight to the workings of the office; Nancy Mayberry searched through her considerable graphics holdings to help us illustrate the report; and Sandra Brown assisted with our efforts in the district library. Charles Camillo, historian at the Mississippi Valley Division, also provided valuable assistance. I am grateful to the Corps of Engineers History Office for granting me permission to publish this work.

A group of diligent researchers assembled many of the pertinent records. Most notably, Matt Schandler undertook the arduous chore of fighting through the post-storm traffic snarls to review project files at the district office and assisted with archival research. Bryan Landry and Jennifer Melancon Cook both did essential work on the bibliography and chronology, plus undertook important archival field work. Additional archival assistance came from Kathryn Norseth, a former colleague from my days in Maryland, Cary Beshel in Fort Worth, and Caree Banton in New Orleans.

Archivists at both the National Archives and its Records Centers also assisted the researchers. Amy Sumpter capably assisted with a second phase of historical research in 2007. Clifford Duplechin efficiently transformed my sketches into readable maps, and DeWitt Braud provided the digital imagery for the first illustration.

A troop of retired Corps employees agreed to sit for interviews and, in doing so, provided incomparable insight into the development of this system. My special thanks go out to them.

I am also deeply indebted to the library staffs at the Hill Memorial Library at Louisiana State University, the Louisiana Collection at the University of New Orleans, the Special Collections at Tulane University, and the New Orleans Public Library. The librarians based in New Orleans deserve special thanks since they were dealing with massive disruptions caused by Katrina. Staffs with the Jefferson Parish Police Jury and their drainage and levee districts and the Orleans Levee District also assisted with access to records, sometimes in extremely difficulty situations, that aided this investigation. Staffs at both the New Orleans and Jefferson Parish drainage authorities also provided access to essential documents.

Many investigations have explained the circumstances that produced flooding in New Orleans during Katrina. I have purposefully limited my use of these post-Katrina sources in order to maintain an independent perspective. Nonetheless, I do cite the key reports in the conclusions.

Also, I am grateful to Craig Gill at the University Press of Mississippi for his efforts to see this project into print and to Tammy Rastoder for her thorough and capable editorial attention to the manuscript.

I am deeply indebted to these individuals and many others who assisted along the way.

PERILOUS PLACE, POWERFUL STORMS

■ ■ ■ ■

INTRODUCTION

Hurricanes have been a constant, if irregular, threat to New Orleans and its vicinity since the city's founding in 1718. Even before surveyors platted the old Vieux Carré, a hurricane swept over the incipient settlement. Back-to-back storms in September 1722 and 1723 destroyed much of the new colonial capital, and at least nine additional storms battered the city before the Louisiana Purchase in 1803. Since becoming part of the United States, another thirty-seven storms have lashed the city, and others have swept ashore nearby, delivering rain and storm surges to Louisiana's coastal wetlands and Mississippi's beaches.[1] New Orleans and its residents are no strangers to the power and damage delivered by hurricanes.

The impact, although not the power, of the most notorious tropical storm to roar across New Orleans and its environs exceeded that of previous events. Despite both familiarity with the threat and recent experience, public officials and citizens were unprepared for Katrina's overwhelming devastation in 2005. Levee failures allowed waters from Lake Pontchartrain and the Gulf of Mexico to pour into much of the city and neighboring parishes. Then, held captive by the ring of levees, the waters inundated as much as 80 percent of the city proper—from a few inches to upwards of fifteen feet. Hurricane-protection levees withstood the storm in neighboring Jefferson Parish, but the massive surge overtopped levees in St. Bernard and Plaquemines parishes, and swept across Grand Isle. Damage to thousands of houses and businesses disrupted the economy of the city, the state, and the region. Additionally, hundreds of thousands of residents were displaced for months, tens of thousands for years. Despite both structural systems and social organizations built to minimize hurricane impacts, Katrina overwhelmed them all. The tragic impacts were not just the result of a short-term meteorological event but the outcome of flawed and incomplete human preparations. Those preparations represent protracted developments and the involvement of many organizations over decades. At the center of the preparations was the set of hurricane-protection levees largely designed and built by the U.S. Army Corps of Engineers and its multiple local partners. Only a long-term, historical account can expose the complex interplay of the numerous players in the failed effort to protect New Orleans from tropical storms.

▪ ▪ ▪ ▪

In the nineteenth century, the Corps of Engineers was a reluctant participant in flood protection. During the early 1800s, it earned an enviable reputation for hydraulic engineering through its efforts to maintain the country's navigable waterways. Charged with pulling snags from internal waterways and designing and installing works that kept the Mississippi and other rivers open to commerce, the Corps compiled an impressive resume. Its programs transformed the wild and untamed internal rivers into what environmental historian Richard White would call "organic machines"—natural systems reworked to serve human needs. Snag boats pulled navigation hazards from channels, and crews removed massive rafts from rivers to open then to steamboats. Wing dams put the river to work maintaining its own channel, and channel straightening reduced the length of meandering streams. Such constructions led historian Todd Shallat to label the Corps as a "nation builder." That is, its efforts aided the expansion of the American republic by modifying waterways to maintain effective linkages among the country's far-flung regions. In this regard, it became known as the foremost engineering organization in the country, but with responsibilities focused on navigation, not flood protection.[2] Its transformation of riverways expanded in the twentieth century. Dams helped regulate flow and maintain adequate depth for shippers, along with reducing flooding, while locks enabled towboats and barges to move up and down rivers reconfigured into hydraulic stairways.

When Congress created the Mississippi River Commission in 1879, it broadened the Corps' responsibilities to include designing levees for the lower Mississippi River. The new commission had an exceptionally tight relationship with the Corps and eased the army engineers into its current flood-protection responsibilities. Congress specified that the president would appoint a seven-person commission, three of them being from the Corps and one of the Corps' representatives serving as commission president. The Corps would provide the engineering and hydrologic expertise for the commission's projects. Initially, Congress appropriated funds specifically for navigation improvements, *not* flood protection. Following a vigorous debate, the commission adopted the theory that a river confined by levees would scour a deeper channel and thus levees would serve navigation. Floods along the lower river following the commission's formation called this policy into question, however, and ultimately forced a policy adjustment. Congress passed the Flood Control Act of 1917, which authorized levees for flood protection for the first time.[3] After the disastrous flood of 1927, Congress enacted a new flood control act that produced a much more prominent role for flood-protection levees along the river and modified the levees-only policy with one embracing a "levees and outlets" approach. This new policy sought to restore drainage ways to the Gulf of Mexico that could divert a portion of flood waters much as the distributary bayous had done before the construction of the river levee system.[4] A few years later, the 1936 Flood Control Act clearly advanced the notion that flood protection was a proper federal activity for all susceptible locations. It explicitly sought to protect residents of floodplains and to provide assistance following inundations. Before the mid-twentieth century, flood protection had become the primary purpose of the levee-building program, and the Corps was thoroughly involved. Hurricane-protection levees had yet to emerge as a priority, but they were on the horizon for Louisiana.[5] The great concrete seawall built in Galveston, Texas, after the devastating hurricane there provided a working model.

Despite the shift from a "levees only" to a "levees and outlets" policy, the sinuous lines of earthen barriers remained the most basic tool in the Corps' flood fighting tool kit. The flood of 1927 in particular offered two contrasting lessons for the Corps and residents of the lower river valley. The Corps adjusted its policy after accepting the futility of levees as the sole defense against high water. But unlike their counterparts in the Mississippi Delta above Vicksburg, residents of the Louisiana delta region saw the levees hold. The only breach was intentional—when the Corps blew a hole in the levee at Caernarvon to reduce the risk of levee failure at New Orleans. That event, along with nearly four decades of protection from river floods, reinforced a steadfast faith in levees. There are few things more impressive than standing in the French Quarter watching oceangoing ships pass behind the levees, seemingly floating above street level.

Levees, from the French term *lever* (to lift), are a centuries-old technology that arrived in the lower Mississippi River with the early French settlers. Borrowing soil from the *batture* or river bank, settlers, and their enslaved workers, would construct linear earthen mounds parallel to the river.[6] Levees are massive and heavy and consume considerable real estate. For every foot of height, levees require five to six feet of horizontal base. A six-foot-tall levee tapers in from its thirty-to-thirty-six-foot base. Constructed of locally available soils, builders sought sufficient clay to bind the levees and minimize movement of water through the structure. For centuries, levees, by many different names, had provided protection from river flooding and raging seas. Selection of levees for flood protection is a considerable commitment. Not only are they costly to build and maintain, they consume land that might be put to other uses. Furthermore, they commonly force a path dependence on their builders. Once the investment is made, it becomes untenable to remove them and forge a different path. Granted, modifications can be incorporated—such as the levees and outlets option—but seldom are these systems abandoned or dismantled. One fundamental reason is that levees encourage development in previously flood prone areas, and once in place, the costs of flooding in the absence of barriers would leap well above the pre-levee costs.

Before the erection of the New Orleans-area hurricane-protection levees, geographer Gilbert White and his students began to question the strict reliance on structures, such as levees, for flood protection. A fundamental concept advanced by White was that structures provided a false sense of security and actually drew development into harm's way. He argued that "the Corps of Engineers is put in the position of exercising high skill in design of protection works while further encroachment is fostered by other activities. It is an army resolutely pushing back an enemy on one frontier while he infiltrates the territory from other frontiers over which it exercises no control."[7] This conflicting responsibility presented a tremendous challenge and an escalating responsibility as flood protection fostered development and development demanded greater flood protection. Since levees and other structural devices used to guard against floods have a design limit, they do not guard against the most extreme storms. To balance costs against benefits, engineers design levees to guard against the most common events that would be expected at a location but concede the floodplain to the rare flood event. Consequently, when an extreme event overwhelms the structures, calamity follows. Thus, disaster is a function of social actions as much as meteorology. To avoid calamity, White advocated restrictions on development in flood-prone areas. Yet, land-use regulation was beyond the sphere of Corps authority, residing with local governments that chronically have taken insufficient action.[8]

In recent years, environmental historians have explored the role of human society in contributing to the disastrous outcome of extreme floods. While most do not acknowledge White's contributions, the notion that people put themselves in harm's way, thereby transforming extreme events into human disasters has great currency in explaining the history of hazards. Ted Steinberg eloquently confronts the notion that by occupying floodplains society courts disaster and by passing extreme events off as "acts of God" it has denied its own culpability in creating calamities.[9] Mike Davis and Jared Orsi both have traced the perils produced by urban expansion in the Los Angeles area.[10] Ari Kelman uses Steinberg's terminology—"Act of God"—to discuss how New Orleans dealt with the flood of 1927. Elsewhere, I have traced the expansion of the city into ill-suited wetland environments and the consequent need repeatedly to enlarge the structural flood-protection system.[11] Stephane Castonguay has argued that, behind barriers, society loses its ability to prepare for extreme events—a process known as preparing for the "regular" among hazards researchers.[12] The common theme found in environmental history and hazards research—that structures provide a false sense of security and compound the flood risk by luring development into flood-prone areas—aligns with White's.

White did not just offer a critique of the structural approach. He recommended a revised plan that came into existence with the passage of the

1968 National Flood Insurance Act. A core plank in White's recommendation was a fundamental realignment in flood-protection policy—a shift from structural to land use. That is, local governments should take steps to restrict the development of flood-prone zones. The policy, which came into law partly in response to heavy flood damages sustained in New Orleans owing to Hurricane Betsy in 1965, seeks to transfer the cost of risk from the population at large to those who live in flood zones. It does this by making available flood insurance to floodplain residents, thereby placing the costs on those who will seek compensation after a flood. The act also promotes local actions to limit floodplain development. Local governments must meet minimum standards for flood mitigation in order for their citizens to have access to the federally underwritten insurance. Despite the program's ambitious goals, enrollment has never been high in most of the country, and development continues to encroach on flood zones. In New Orleans, where the entire city is on floodplain[13] and many structures predate the policy, the flood insurance program has had modest impact on land-use practices. Yet, New Orleans and vicinity has one of the highest flood insurance subscriber rates in the country—57 percent before Katrina. Residents have come to regard this insurance as an essential buffer against financial loss. But local government did little to divert development from areas that had suffered from hurricane-induced flooding.[14]

Although oriented to river flooding, the national flood insurance policy also applies to hurricane-induced flooding. Congressional support for the program, in fact, grew following Hurricane Betsy, which flooded parts of New Orleans in 1965. But a finely tuned federal focus on hurricanes began a decade before. Coastal development along the Atlantic seaboard and the Gulf Coast had put more property at risk in the first half of the twentieth century. Following a series of powerful storms that produced huge loses along those developed coasts in the 1950s, the federal government produced a series of reports on hurricanes. The research efforts sought to improve hurricane forecasting, measuring the impacts of storm surge, and preparation for storms along the Atlantic and Gulf coasts. The investigation had little to say about structural protections. Indeed, the Model Hurricane Plan for a Coastal Community emphasized evacuation, not hardened structural defenses.[15]

Nevertheless, those who lived in New Orleans annually had witnessed the benefits of federally constructed river levees. The economic disruption produced by hurricanes and displacement of Galveston as Texas's major city after the 1900 hurricane was well known.[16] And although evacuation also had helped coastal residents survive storms, leaders and residents in southeast Louisiana sought more tangible protection.

In the New Orleans vicinity, local organizations had built modest barriers to protect against storm surge and wave along the lakefront and along the Industrial Canal by mid-century. Following the hurricane of 1947 that inundated large sections of the lakefront in both Jefferson and Orleans parishes, local interests called on the Corps to shore up the levees along the lakefront. As part of the larger national focus on hurricanes during the 1950s, the Corps commenced planning for structural hurricane-protection projects at a national level. In the New Orleans District, this entailed initial planning for hurricane-protection works for the Lake Pontchartrain area.[17] More than any storm up to the mid-1960s, Betsy propelled local congressional and public support for hurricane protection to the national level, and appropriations began to flow into a multicomponent structural system for the New Orleans area. While never rivaling river levees in cost or scope nationally, hurricane protection became a sizable task of the New Orleans District Corps of Engineers after 1965. Planning and construction for this work, while linked to the river flood-protection system, was largely a separate activity and employed a lower standard of protection.

Several distinct hurricane-protection units emerged from the planning efforts of the New Orleans district office (see fig. 1.1). First, the Lake Pontchartrain and Vicinity Project sought to protect the portions of both Orleans and Jefferson parishes that faced the lake. It also included the portions of St. Bernard Parish immediately downstream from New Orleans. Several separate levee systems partially encircled urbanized areas and tied into the river levees to complete the enclosure around these tracts. Another levee system completely encircled eastern New Orleans and did not tie into the river levees. Neighborhoods of New Orleans and St. Bernard Parish downstream from the Industrial Canal also received enlarged "back" levees that connected to the river levees. The second component to the southeast Louisiana hurricane-protection system was the New Orleans to Venice

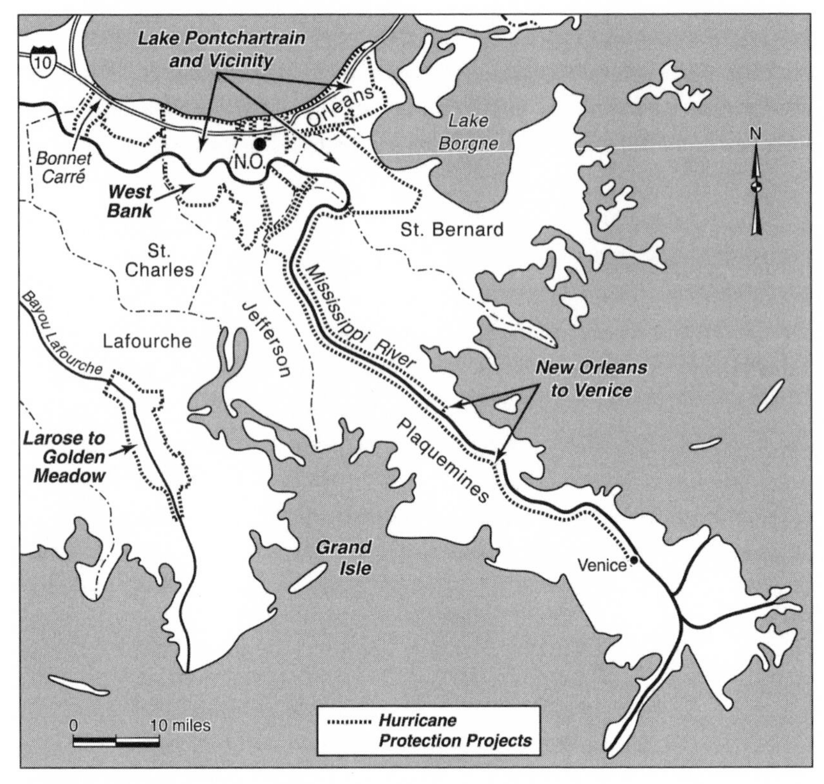

FIGURE 1.1 Hurricane-protection projects in southeast Louisiana

project. It protected lower St. Bernard and Plaquemines parishes with back levees linked to river levees. The West Bank project was the third component and offered protection for portions of Jefferson Parish on the West Bank of the Mississippi River by erecting a series of back levees that also joined preexisting river levees. Along lower Bayou Lafourche, the Larose to Golden Meadow project produced a ring levee around the natural levee of the bayou and a pair of floodgates to permit the closing of the waterway during storms. The final component was the Grand Isle project. It included a series of barrier dune enhancements and structures to control erosion.

As in the case of riverine floodplains, structural protection encouraged development in flood-prone areas. Orleans, Jefferson, and St. Bernard parishes witnessed extensive development within their rings of hurricane-protection levees.[18] Likewise, development in Plaquemines and Lafourche parishes and on Grand Isle also continued once protection was under construction or completed. Contrary to the land-use ideals of the National Flood Insurance Program, suburbs erupted in low-lying areas encircled by the massive earthen barriers. Levees designed to protect encouraged inappropriate floodplain development.

Construction of structures to protect the lower river parishes from tropical storms proceeded slowly. Lawsuits over environmental issues forced adjustments in planning and design. Local interests sought to take over portions of the process to maximize economic development potential. Throughout the interval between Betsy and Katrina, Corps officials encountered situations that delayed construction progress and kept New Orleans exposed to hurricane risks. Despite years of proclamations that one of the greatest potential disasters facing the country was a major hurricane strike in New Orleans, the hurricane protection system remained a work in progress when Katrina made landfall in late August 2005. The intense sense of urgency that propelled actions in the years immediately following Betsy eroded over the decades, but that is no longer the case.[19]

In the next chapter, I examine the local environmental context that makes New Orleans and the lower delta susceptible to hurricanes and consider the city's experience with hurricanes and federal policy prior to 1957. In subsequent chapters, I look at the period from 1957 to 1969, when a spate of powerful storms prompted an overhaul of federal policy and pushed the New Orleans District into a prominent role on hurricane protection engineering. The slow process of implementing the Lake Pontchartrain portion of the system and the complex interaction between the Corps and local interests will be the subject of the subsequent two chapters. Hurricanes and efforts to complete the hurricane protection system since 1990 will provide the focus for the following chapter.

A project begun in earnest in 1965 was still not complete in 2005. This situation was not desired by the local interests or the Corps but was the result of a complex set of environmental, social, political, and financial issues. I will attempt to sort out the role of the many actors in this drama and discuss the sequence of events prior to Katrina that contributed to the human disaster.

■ ■

CITY AT RISK

New Orleans occupies a place on a subsiding deltaic floodplain, on land built and formerly sustained by Mississippi River floods. Residents of the city and neighboring agriculturalists shared a fundamental concern with keeping the flood waters out of the city and off farm lands. Colonial policy, and subsequent state policy, sought to maintain structural defenses, at the same time deferring flood protection costs to other entities. Colonial law placed the levee-building responsibility on individual land owners, although local levee boards assumed that duty after statehood. By the late nineteenth century, Congress tasked the Mississippi River Commission with levee building and thereby shifted much of the financial burden to the federal treasury. Although flood protection until that time had remained focused on river threats, transferring that responsibility to the national level opened the door to federal obligations for hurricane-generated inundation as well. In this chapter, I examine the evolution of flood-protection authority in the lower Mississippi River Valley, particularly with respect to New Orleans. I will trace the impacts of late–nineteenth- and early-twentieth-century hurricanes and examine local preparations for tropical storms. As with river flooding, responsibility began with individuals or families and gradually progressed to the federal level. The local setting of New Orleans provides unique challenges in light of cyclonic storms. As a preface to the extended discussion of federal projects, I will consider both the physical features of the region and pre-1957 efforts to provide erect hurricane protection.

LAY OF THE LAND

New Orleans and its suburban neighbors occupy a site of exceptional vulnerability. As part of an active delta and floodplain, the dynamic Mississippi River system literally created the site by layering deposits of sediment each spring when the river escaped its banks. Early human settlement took advantage of the subtle variations in elevation produced by the river's land-building processes.

Native Americans had valued the site of New Orleans for its portage advantages, linking the Mississippi River and Lake Pontchartrain and its tributaries and the Gulf of Mexico. Indigenous people introduced the French to this location, which had geopolitical, economic, and flood-protection value to Europeans. The favored terrain where the French established New Orleans had always been the higher ground known as the natural levee. Along either side of the river stand shoulders of slightly elevated land. Elevations of about twelve feet above sea level exist adjacent to the river in the vicinity of New Orleans. Flood waters carried sediments in suspension, and when they escaped the channel, they lost velocity and dumped the coarsest sediments, such as sand, near the river creating the natural levees. Finer-grained clays stayed in suspension longer and settled out some distance from the river. This natural sorting process created a gentle slope from the river toward the backswamp and the lakefront (see fig. 2.1). The natural levee is about one and a half miles wide from its crest to its toe on either side of the river at New Orleans. Moving away from the river, elevations decline to below sea level in the interior of the crescent, which gives the city its most prominent nickname—the Crescent City. Human settlement, even before the French arrived, logically clustered on the higher natural levee for several reasons. It was the last to flood and the first to drain; and the sandy soils drained water more effectively and hence proved superior for cultivation and road building.[1] Unlike landscapes created by rivers that incise themselves into bedrock and slope upward as one moves away from the channel, the lower Mississippi River floodplain was an inverted topography. Proximity to the river for transportation and for flood protection fostered elongated European settlements that took advantage of the modest but critical natural levee.

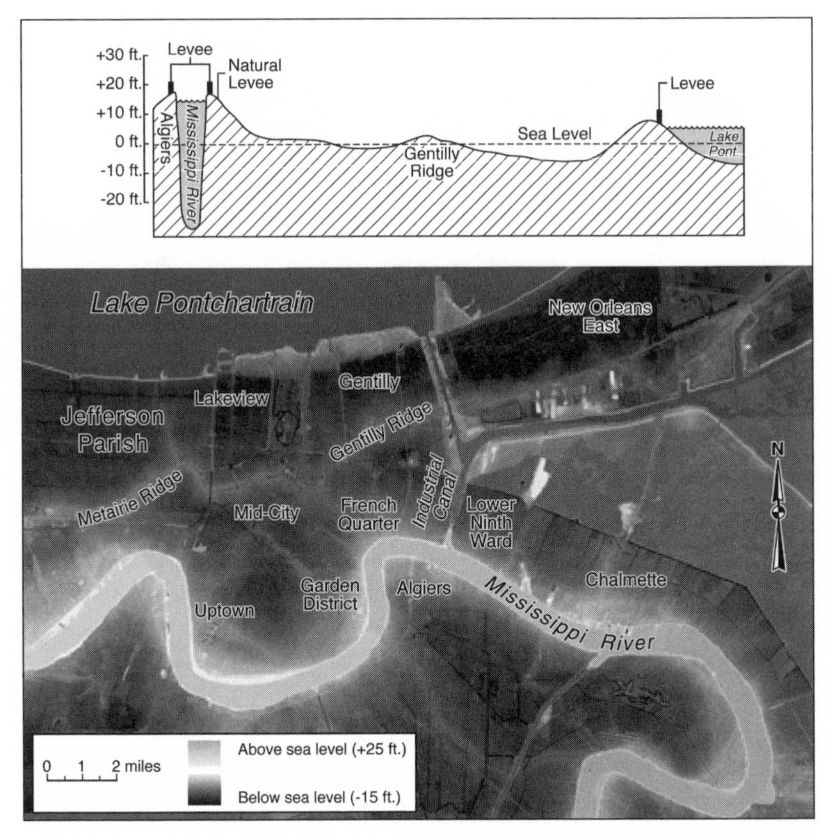

FIGURE 2.1 New Orleans topography and cross section. Mosaic of New Orleans created by DeWitt Braud, LSU. Based on USGS Digital Elevation Model (DEM) derived from Lidar elevation data funded by FEMA and the Corps of Engineers, and flown by 3001, Inc., 2003.

Snaking across Jefferson and Orleans parishes between the river and Lake Pontchartrain is another slight rise. Known as the Metairie Ridge in its western section and Gentilly Ridge to the east, it is a relict natural levee left by an ancient course of the Mississippi River abandoned long before the French arrived. This ridge offered a secondary zone for activities that demanded relatively higher and better drained soils: land transportation, residential and agricultural land uses, and ultimately cemeteries. This older natural levee is about one mile wide near its western extreme in Kenner and tapers off east of the city to the point it loses all surface expression. Maximum height of the relict natural levee is seven feet, although five-foot heights are common in New Orleans. This slight rise provided limited protection from storm surge from Lake Pontchartrain for the early settlements that clustered on the current river's natural levee.[2]

Bayou St. John, a natural waterway, had carved a channel through the ridge and provided natural drainage for the city. This waterway played an important role in the city's early settlement but also posed a hazard. It offered Europeans backdoor access to the city, which solidified New Orleans's strategic position as a political and economic center in France's New World empire. There were costs associated with this situation as well. A low area in the center of the great crescent of natural levees was at or below sea level (see fig. 2.1). With so little gradient between the natural levee and the mouth of Bayou St. John, winds could drive storm surge up this channel and did repeatedly in the nineteenth century.[3] During the great river floods of the nineteenth century, waters collected here at the "bottom of the bowl." When sufficient water accumulated, a portion would drain into the lake via Bayou St. John.

Lakeward of the ridges stood an expansive cypress forest at elevations about a foot above sea level that graded to marsh as the land surface approached sea level. As the city grew lakeward, entrepreneurs constructed ring levees and pumped water out of the wetlands in order to create new real estate. Yet, peaty soils beneath the swath of land near the lakefront compress and oxidize after they are drained. Once levees prevent immediate resubmersion and mechanical devices remove surface and groundwater, the land sinks. This subsidence has contributed significantly to flood damage following recent hurricanes when flood waters overtop or breach the levees and collect within the now below-sea-level terrain. Buried relict sand beaches, which are not subject to subsidence, underlie portions of the lakefront and provide more stable footing for construction and slightly higher ground.[4]

Downstream from New Orleans are St. Bernard and Plaquemines parishes. St. Bernard's population and industry are concentrated along the natural levee. A second linear cluster of human settlement follows the smaller natural levee of the Bayou Terre-Aux-Boeuf, which is surrounded by marshlands built centuries ago as part of the ancient St. Bernard delta. Plaquemines Parish is a narrow strip of land on either side of the river as it passes into the river's bird-foot delta. Land levels are only a few feet above sea level on the highest natural grade. Adjacent to the natural levee are wetlands. The

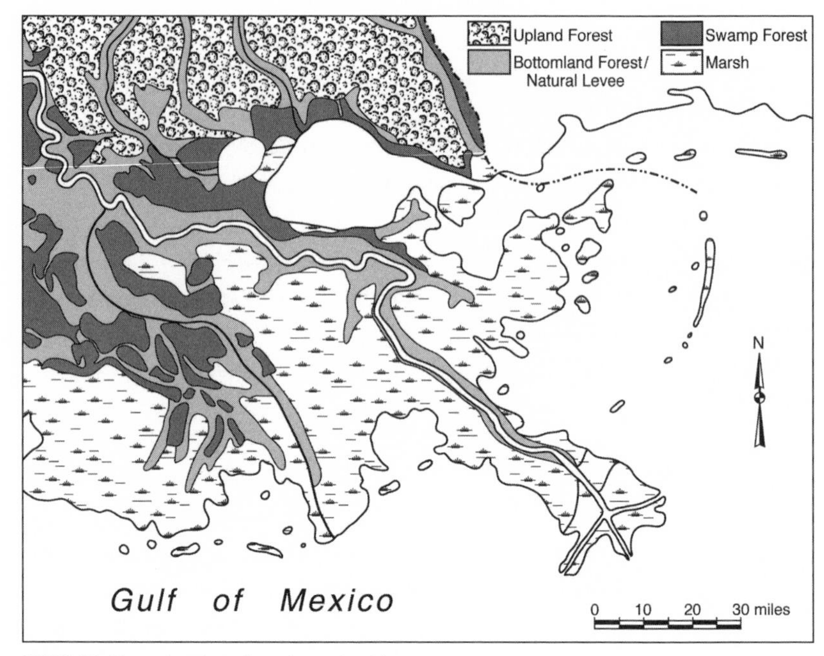

FIGURE 2.2 Natural setting of southeast Louisiana

marshes in both these parishes have been subject to subsidence in the years since humans flanked the river with levees, inhibiting riverine flooding and starving these precarious lands of rejuvenating sediments. Canals cut through the wetland and oil and natural gas extraction further compound the loss of these vital nurseries and surge buffers.

Lake Pontchartrain itself is also a critical component in the city's susceptibility to storm surge. Although named a lake, it is a brackish bay of the Gulf of Mexico. Two natural channels link the lake to the gulf. The Rigolets and the Chef Menteur Pass allow the movement of tidal flow and saltwater into Lake Pontchartrain. A low marshy peninsula separates most of the lake from Mississippi Sound to the east and Lake Borgne to the southeast. A series of streams drain the Pleistocene terrace on the lake's northern flank and are the source of considerable fresh water. Cypress-tupelo swamps border the lake's western shore and serve as a buffer between the brackish water body and the freshwater Lake Maurepas (see fig. 2.2). Approximately twenty-five miles wide on its north-south axis and forty-one miles from its western shore to the Rigolets, Lake Pontchartrain is a sizable body of water.

Its shallow depth (averaging twelve feet), size (640 square miles), and relative location make it a source of considerable storm surge potential.

New Orleans's topographic condition has rendered it extremely susceptible to both river floods and hurricane storm surge over the years. Peirce Lewis characterized the city as an impossible but inevitable city. "Impossible" because of its inauspicious physical traits, susceptible to floods and hurricanes. "Inevitable" given its strategic position near the mouth of the major river of North America and an entrêpot for commerce.[5] Swamps and marshes had to be drained to make the area inhabitable. Once that occurred, subsidence accentuated the city's susceptibility to floodwaters.

LEVEES AND HURRICANES

Flood protection involved more than just levees (although levees were the most visually prominent and least flexible element) and started at home. Although French architectural practices produced what became known as *poteaux en terre*—or cabins with vertical cypress posts driven into the ground with dirt floors. Builders on the delta soon introduced Creole construction practices from the Caribbean, in particular, raised houses, set on piers some five to six feet above the ground. Used in the Caribbean to capture breezes and elevate the living quarters above the most oppressive mosquito zone, elevated houses served the additional purpose of providing some protection from floods. The combination of raised houses, built on the natural levee, reduced the impact of almost annual inundations.

For additional protection, colonial-era settlers commenced the construction of massive earthworks along the river. To protect the outpost known as the French capital, workmen heaped up the delta soil and shaped it into ridges about four feet high between the city and the river. By 1727, they had completed about a mile of levee. This supplemented the architectural devices for those who could afford to build above the land surface. Given local topography, floods that escaped the channel well upstream from the incipient city could still produce floods that would encircle the town and farms from the rear. To offset the "backdoor" flooding, a colonial law passed in 1728 required rural landowners to construct levees between their

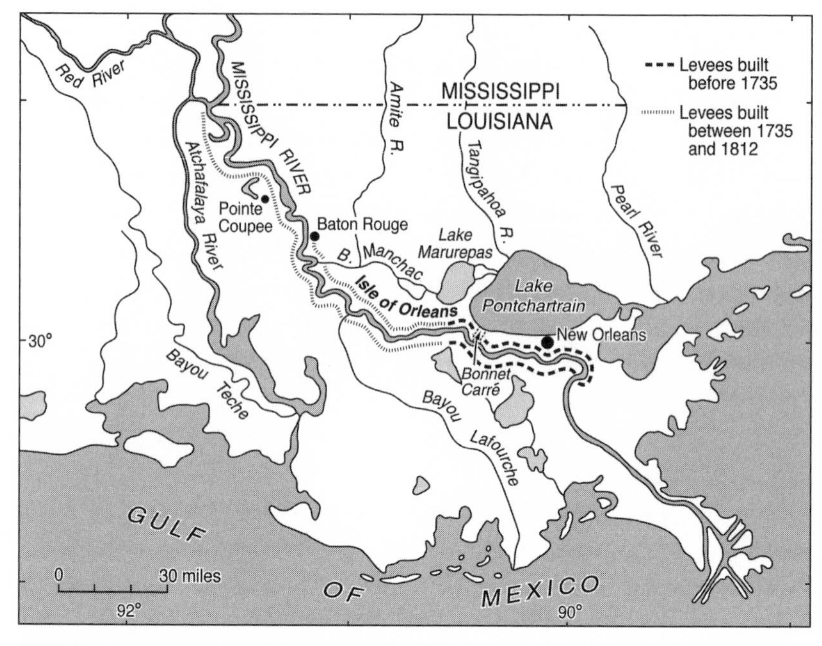

FIGURE 2.3 Levees built by 1912. After Cowdrey 1977.

fields and the river. Consequently, by 1735, levees extended tens of miles up and down both banks of the river (see fig. 2.3).[6] Privately-built levees extended the protection for New Orleans far beyond what the city could have undertaken or the colonial government could have afforded. Only by extending the levees far upriver could officials offer even rudimentary protection to New Orleans.[7] Spanish officials largely continued the policy of private levee building between 1763 and 1803, as did the territorial government of the United States after the Louisiana Purchase. By the time of statehood (1812), landowners had constructed levees upriver to Baton Rouge on the East Bank and nearly to the junction with the Red River on the West Bank. Despite their impressive appearance, the levees were a weak barrier against spring floods.[8]

Even with levees lining the river, high water crept into the city from the backswamp in 1791 and 1799 during the late Spanish period, and, in rapid succession, floods threatened the city in 1809, 1811, 1813, and 1815. A huge flood caused considerable disruption to New Orleans in 1816 and covered parts of the city with water for up to a month. This flood, which broke through the levees upstream from the city, made evident the fact that the city's bulwarks were superior to those built and maintained by individual landowners. As a system, however, the barrier was far from satisfactory. Extraordinary flood crests on both the Red River and the Mississippi River in 1823 produced many crevasses between Baton Rouge and New Orleans during the early summer. In early July, the levees failed above the city, and high water entered the city, where it remained for at least two weeks. The flood of 1844 caused massive flooding upstream from New Orleans when a crevasse broke through the poorly-made levees at Bonnet Carré.[9]

The Sauvé Crevasse of 1849 reemphasized this condition. A sizable crevasse or break in the levee occurred seventeen miles upstream from the city during the spring rise. Water poured through the gap and turned toward the city when it encountered the Metairie Ridge that paralleled the river. This slight topographic rise directed the water into the lowest sections of the city where accumulations reached a depth of six feet. As efforts to plug the crevasse failed, water filled the "bottom of the bowl" and eventually crept into the Vieux Carré. Damage was most pronounced in the low-lying quarters, generally occupied by recent immigrants who were also the city's poor. Floodwaters forced some twelve thousand residents from their homes and invaded about two thousand structures. High water disrupted commerce for forty days before efforts to drain the floodwaters into Lake Pontchartrain finally succeeded.[10]

Federal assistance to the local levee-building projects came with the passage of the Swamp Land Acts of 1849 and 1850. These two acts transferred some 1.2 million acres of federal swampland to Louisiana. Conditions of the transfer dedicated proceeds from the sale of these lands to levee construction. This policy adjustment maintained local control of levees but acknowledged the fact that protecting the lower river, particularly the commercial *entrepôt* of the Mississippi River valley, benefited the entire nation. This indirect federal subsidy acknowledged the national dimensions of flood control on the lower river.[11]

By the 1870s, the local and state efforts had raised the levees at New Orleans to a sufficient height and reinforced them to a sufficient strength to be able to withstand several serious floods that overwhelmed the upstream levees and caused extensive damage in rural agricultural districts. High water in 1874 that produced extensive flooding, while not impacting New

Orleans, prompted the federal government to consider more direct participation in flood control. Following a study of the situation, federal engineers recommended improvements, but no action ensued. A second assessment in 1879 produced more tangible results. An engineering report argued that levees along the Mississippi would enhance river navigation, the Corps of Engineers' traditional responsibility, and Congress subsequently passed legislation creating the Mississippi River Commission. This new body had the mission to design an effective levee system for navigational purposes. But before Congress funded levee construction, floods swept through the valley in 1881 and 1882, causing extensive damage. After the second and more devastating of these events, Congress appropriated funds to begin the gradual reinforcement of the rural levees from Cairo, Illinois, to the mouth of the Mississippi River.[12]

Even with levee improvements, New Orleans remained vulnerable to "backdoor" flooding throughout the nineteenth century. River floods in the 1870s broke through the levees and poured into Lake Pontchartrain. It rose to such a height that it invaded the city's rear districts. Again in 1890, upstream crevasses caused the lake to overflow into the city. Water rose two feet deep in houses along the lakefront, drowned roads and railroads, and flowed up the drainage canals, invading adjacent neighborhoods normally secure from flooding.[13] Creation of the Mississippi River Commission and funding much of the levee building with federal funds began to diminish individual involvement in flood preparation. No longer were individual landowners responsible. Although exceptionally high water tended to mobilize entire communities, such events became rare, and communities lost concern with regular floods hidden from view by the growing levees.

Protection from hurricanes followed a similar path, and individual protection constituted the first response. Hurricane storm surge from the lake, not the river or the gulf, presented the greatest threat to New Orleans, but the natural levee still provided the first level of protection. In addition, by building raised structures on the natural levee, homeowners enjoyed a limited amount of protection from the relict natural levees—the Gentilly and Metairie ridges. Urban residents also developed resorts on barriers islands where they could seek refuge from the summer disease outbreaks. Two calamitous hurricanes wiped out nineteenth-century coastal resorts and killed hundreds in each situation. Isle Dernier ceased to function as a resort after the 1856 hurricane, and the resort hotel never reopened on Grand Isle after the 1893 storm. These actions represent another means of seeking protection from storms: retreat from highly exposed locations. Despite abandoning coastal resorts, the city continued to grow toward its most susceptible exposure—the Lake Pontchartrain shore. Hurricane storm surge in 1887 concentrated damages in the rear of the city and along drainage canals that backed up with lake water. Storms battered the New Orleans lakefront and caused urban flooding in 1901, 1905, and again in 1909.

These early-twentieth-century storms reveal how the region dealt with tropical storms before it began to build flood-protection structures. Local officials received warnings from the Weather Bureau about approaching storms, although imprecise by today's standards, so there was a degree of anticipation and short-term warnings. Shipping companies would order ship captains to anchor in the river and ride out the storm inland, rather than venturing into the Gulf of Mexico. Railroad companies would load flatcars with ties and rail and put crews on standby to repair breaks in their lines as swiftly as possible. Public utility and communication companies also had crews prepared to promptly repair service disruptions after storms passed. Once the Weather Bureau posted warning flags in the coastal areas, fishermen and residents of the low-lying marshlands could secure their equipment, and some evacuated to New Orleans. City officials warned residents and visitors to remain indoors in the city. While there were fatalities, loss of life was seldom large—other than the resort tragedies—and New Orleans served as the regional refuge. With a more substantial wetland buffer surrounding the city, many raised houses and little residential or commercial development on the lowest areas, the city was far safer than the coastal margins. Both the city and the National Guard provided disaster assistance after the storm, but preparations and protection before landfall remained largely a personal/corporate responsibility.[14]

A massive hurricane struck in 1915 and caused extensive damage when surge and waves attacked the vulnerable lakefront. With little more than modest drainage district levees lining parts of the shoreline (see fig. 2.4), there was little human-made protection from high water. The Gentilly and Metairie ridges offered some protection, but several openings cut through these natural barriers. Floodwaters rose as high as eight feet in the city owing to the combined impact of heavy rains and storm surge pushing

FIGURE 2.4 Lakefront seawall at Milneburg, 1927. Seawall provided only modest protection against storm surge. Courtesy Louisiana Division/City Archives, New Orleans Public Library.

up the various drainage canals and Bayou St. John. It took several hours for the city's pumps to transfer the floodwater back to the lake. Wind and surge destroyed up to 90 percent of structures along the lakefront and in communities downstream from New Orleans.[15] This terrible storm prompted a shift from personal to community hurricane protection. During the 1920s, the Orleans Levee Board began construction of a seawall along the lakefront. A sinuous, stepped concrete wall rose nine and a half feet from the water several hundred feet out from the natural shoreline (see fig. 2.5). Plans called for the movement of lake-bed sediments to the area between the former shoreline and the new seawall, creating an extensive tract of "made land" completed by 1934. Although it was not a true hurricane protection levee, the seawall designers sought to erect a structure that would have kept the 1915 storm out of the city. In neighboring Jefferson Parish, the Louisiana Highway Department had constructed a road embankment along the lakefront as part of a New Orleans to Hammond

FIGURE 2.5 New Orleans lakefront seawall completed in 1934. Courtesy Louisiana Division/City Archives, New Orleans Public Library.

highway in 1924. It stood from three to six feet high across the width of Jefferson Parish's lakefront and offered a two-to-five-foot barrier in adjacent St. Charles Parish. With the completion of the Bonnet Carré Spillway in the early 1930s, the state shifted the route inland, away from the lakefront. Local levee districts continued to rely on the roadway embankment, but poor maintenance and subsidence lessened the berm's effectiveness as a barrier to hurricane surge.[16]

The New Orleans area passed through a relatively hurricane-free period between the late 1910s and the mid-1940s. And during that respite, the city grew into harm's way. Residential development in the newly protected lands behind the seawall accelerated during the 1920s, paused during the Great Depression and World War II, and was resuming when the next major hurricane hit the city in 1947. An unnamed storm passed over the metropolitan area and propelled winds with gusts up to 112 miles per hour at the airport on September 19. Storm surge easily topped the seawall, and water covered approximately nine square miles of Orleans Parish between the lake and the Gentilly Ridge. Jefferson Parish, protected only by the roadway, suffered even greater damage.[17] The raised highway had subsided in the twenty years since its construction and offered little protection against the wind-driven waves. Floodwaters covered more than 30 square miles of Jefferson Parish for up to two weeks. Subsidence behind the lakefront levees created topographic bowls that held up to six feet of water until the pumps could lift it back into Lake Pontchartrain. In addition, storm surge of up to eleven feet also washed over much of the largely unprotected St. Bernard and Plaquemines parishes.[18]

Renewed levee building, with the assistance of the Corps of Engineers, followed this storm. Congress authorized the Lake Pontchartrain and Vicinity Project to assist with raising the lakefront levee in 1946 and, with additional impetus from the recent storm, enlarged the project's scope in 1950.[19] The Corps began work on an eight-foot-high levee along the Jefferson Parish lakefront to complement the Orleans Parish seawall. Nonetheless, Hurricane Flossy in 1956 exposed shortcomings in still-under-construction barriers. Subsidence of the Orleans Parish seawall left it lower than the original design height, and storm surge overtopped some sections. This allowed much of the Gentilly neighborhood to be inundated. Some eight hundred residents evacuated and a hundred homes suffered damage. High-water marks in the vicinity of three and a half feet were common. Storm surge also caused flooding in the vicinity of the Industrial Canal. Wind-driven water entered the canal from both Lake Pontchartrain and from Lake Borgne via the Intracoastal Wateway, and also overtopped some low sections of the canal levee. Water accumulated to depths of two plus feet in this area. Lakefront areas of Jefferson Parish escaped serious flooding. Storm surge came close to overtopping the Lake Pontchartrain, Louisiana Levee Project, but federal improvements had raised it sufficiently to withstand this storm. Storm surge completely washed over Grand Isle. St. Bernard Parish also suffered extensive flooding and storm related damages. Lower sections of East Bank Plaqeumines Parish also suffered extensive flooding. Those areas with back levees, arcs of levees that looped behind settled areas and tied in to the preexisting river levees, and settlements on the West Bank endured little inundation.[20]

DRAINAGE

New Orleans officials had long seen standing water within the city as a serious problem. During the colonial period and through most of the nineteenth century, medical authorities considered moist ground and decaying vegetation as a source of miasmas, or fumes that contributed to diseases such as yellow fever. Frequent arrival of epidemics during the nineteenth century prompted efforts to drain excess water and thereby alleviate the threat of disease. However, New Orleans's topography presented a challenge for effective drainage. Older parts of the city built on the natural levee enjoyed reasonable drainage, but as the city expanded off the higher ground into the low-lying former swamps, there was little slope to permit gravity to move runoff toward Lake Pontchartrain. In addition, relict ridges stood between the city and the lake and made drainage even more problematic.

Spanish authorities completed the Carondelet Canal in the 1790s, which served the dual purpose of providing navigational access from Bayou St. John and the lake to the rear of the Vieux Carré and also provided some limited drainage benefits (see fig. 2.1). With virtually no current, it quickly filled with sediments and sewage that produced a nuisance to surrounding neighborhoods. Although the accumulated sediments inhibited drainage out of the city, the canal served as a conduit for lake waters driven by north winds that could flood adjacent property. When the city dredged the foul residue from the canal and enlarged its capacity in 1817, it merely created a larger problem.

The city and private entrepreneurs dug additional canals by the 1870s (see fig. 2.1). Three canals devoted principally to drainage connected the city to the lake. The Upper Line (now the 17th Street Canal), the Orleans, and London Avenue canals reached from the Metairie-Gentilly Ridge to Lake Pontchartrain. "Drainage machines" or pumps would lift water from the low ground on the river side of the ridges and allow it to drain lakeward. The Melpomene Canal, with its drainage machine near the bottom of the bowl, directed runoff into the New Canal—an 1830s navigational canal that linked the lake with the American sector of the central business district. Despite these piecemeal improvements, much of New Orleans remained a quagmire during most of the year.[21]

FIGURE 2.6 Peoples Avenue Canal constructed by the Sewerage and Water Board for urban drainage during the early 1900s. Courtesy Louisiana Division/City Archives, New Orleans Public Library.

Progressive Era leaders in New Orleans sought to wrest the city from its soggy terrain once and for all. In 1893, a committee put forward a plan for a systematic and citywide drainage. Although this initial plan never received public approval, it served as the outline for a second attempt to provide water, sewerage, and drainage that won both voter acceptance and public funding in 1899. This ambitious public-works project gradually installed a systematic drainage system. It started with the most populous sectors of the city from 1900 to about 1920, and then extended service to new suburbs toward the lakefront and also some previously neglected sectors. By 1940, an extensive network of canals and pumps had lowered the water table within much of the city (see fig. 2.6).[22]

Draining the city and lowering the water table produced results that adversely affected hurricane vulnerability. After water is removed, peaty soils that underlie much of the lakefront property undergo oxidation and compression. These processes produce subsidence. Consequently, much of the land that has undergone drainage has subsided below lake level. Without levees constructed along the lakefront, they would have disappeared beneath lake waters. Local drainage districts had built the first levees that encircled small tracts that private developers sought to sell to home buyers (see fig. 2.4). Subsequent improvements in response to hurricanes merely accentuated impacts when water washed over the levees.

■ ■ ■

SEASON OF THE STORMS

Coastal Louisiana entered a climatic cycle of more frequent hurricanes for slightly more than a decade following the autumn of 1957. Two massive storms bracket, both temporally and spatially, this period of elevated cyclonic activity for New Orleans: Audrey in 1957 and Camille in 1969. Although neither storm caused extensive damage to New Orleans, the first obliterated portions of the southwest Louisiana coast and the second produced similar devastation to the Mississippi shore. Coupled with tropical cyclones that made more direct hits on the New Orleans metropolitan area, this season of the winds prompted a massive response. Congress assigned the New Orleans District the duty of shoring up hurricane protection for New Orleans. Army engineers considered two possible protection systems, both relying primarily on levees. Controversy over the possible environmental and economic impacts of one design produced a protracted conflict between the Corps and local organizations and ultimately affected both the project's progress and its cost. In this chapter, I will examine the impacts of storms during the 1948–69 period, the Corps' initial response to its congressional mandate, and the ensuing local engagement in the process of supplying storm protection.

HURRICANES, 1948–69

Climatologists had begun studying the cycles of storms related to global climate change by the 1950s. Observations of warmer ocean temperatures

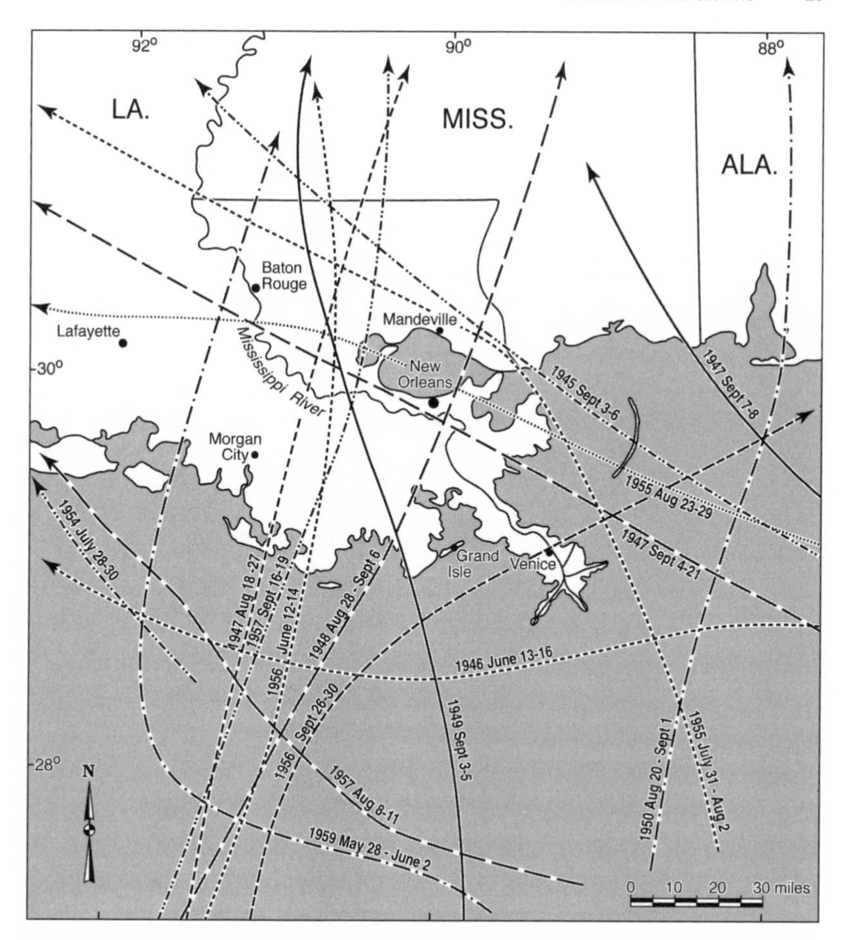

FIGURE 3.1 Hurricane paths, 1945–1960. After U.S. Army Corps of Engineers, 1972.

in the North Atlantic and shifting wind patterns led hurricane experts to conclude that more tropical cyclones were moving onshore in the Gulf of Mexico and the Atlantic seaboard than in previous years and that these storms tended to maintain their intensity longer. Meteorologists speculated that the cycle of more frequent storms striking North America would draw to a close around 1965.[1] Ultimately, the period of more frequent and stronger storms persisted longer than originally anticipated, and subsequent investigations note that the decade from 1961 to 1970 was one of particularly frequent "major" storms in the Gulf of Mexico.[2] Southeast Louisiana and coastal Mississippi experienced some of their most devastating weather events during this spike in tropical storm activity.

Scattered tropical storms buffeted the Louisiana coast following the powerful 1947 storm and before the onslaught of the 1960s (see fig. 3.1). In 1949, a disturbance blew over Grand Isle with winds of 50 miles per hour and produced a storm surge of 4.4 feet at Biloxi, Mississippi, and 4.8 feet of surge at Mandeville on Lake Pontchartrain. Tropical Storm Barbara moved ashore southeast of Lake Charles in 1954 with winds in the vicinity of 60 miles per hour.[3] Neither of these storms produced severe damage. A pair of tropical storms blew across the Mississippi River delta in 1955. The first arrived in late July and generated surges measured at 5.4 feet at Shell Beach, 3.6 feet in Lake Pontchartrain, and 3–6 feet along the Mississippi shore. Once again, with winds in the neighborhood of 60 miles per hour, infrastructure in the gulf and watercraft took the brunt of the damage. The second system swept in from the Caribbean and passed over New Orleans with winds from 40 to 50 miles per hour. Although it piled up a storm surge of 3.6 feet at Shell Beach in St. Bernard Parish, the storm caused modest damage.[4]

In the aftermath of these storms and as part of a national effort to address the need for improved hurricane preparations and protection, the Corps of Engineers held hearings in New Orleans. They solicited input on the impacts of past storms and the need for improved hurricane protection in the New Orleans vicinity in March 1956. Officials from the state and from local levee districts along with private citizens sounded a unified appeal for the Corps' continued, and indeed expanded, participation in hurricane protection. J. J. Holtgrieve from the Jefferson Parish Police Jury reported that since the hurricane of 1947, which had caused extensive flooding to the areas inland from the lakefront roadway, "new dwellings by the thousands have been developed in that area, and any failure in the levee district which has been established along the shores of Lake Pontchartrain will take and endanger the lives and property of untold amounts."[5] His appeal was typical of the concerns relayed to the Corps officials and provided justification for continued planning.

Even more important than public testimony were storms that continued to rake the region. A powerful storm swept over the delta region within months of the hearing. Hurricane Flossy traveled northward from the Yucatan Peninsula, veered toward the east while over the open water of

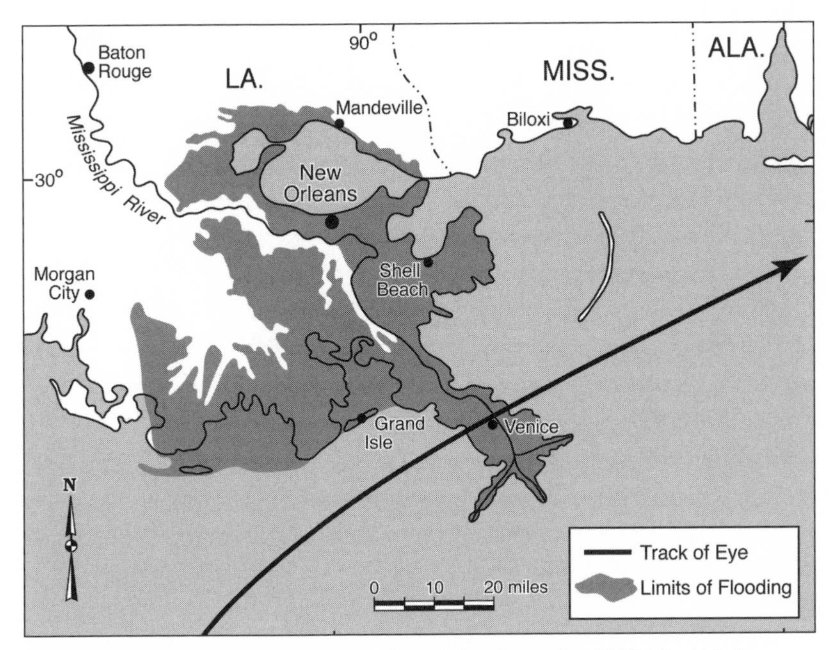

FIGURE 3.2 Hurricane Flossy track and area of inundation, September 1956. After U.S. Army Corps of Engineers, 1972.

the Gulf of Mexico, and tracked across the mouth of the Mississippi River in late September on an east-northeasterly course with winds blowing from 90 to 110 miles an hour (see fig. 3.2). Storm surge ranged from 13 feet at Ostrica Lock to 10.9 feet at Shell Beach. A surge of 8 feet washed ashore on Grand Isle, and the newspaper reported that 300 people had evacuated the recreational island before waves swept over it.[6] Seventy-mile-per-hour winds howled over the New Orleans airport on September 23 and pushed the storm surge over the lakefront seawall and levees and over low areas of the Industrial Canal levee. The Gentilly and Lakeview neighborhoods, near the lakefront, suffered flooding, and some 800 citizens sought refuge in Red Cross shelters. High water flowed inland for a mile and a half to Filmore Avenue and covered the area between Peoples Avenue to the London Avenue Canal. Immediately after the storm, the mayor assured citizens that the city's pumps were dealing with the high water and that they had lowered the water 4.5 feet in the first 12 hours after water overtopped the lakefront barrier.[7] By September 26, most residents of New Orleans had returned to their homes.

Downstream parishes, however, faced a prolonged ordeal. Some 1,500 to 2,000 people fled their St. Bernard Parish homes. Some communities in Plaquemines Parish reported that families lost "everything" to the storm surge.[8] Damage was particularly severe in these lower river parishes. Some 286 square miles of land went beneath water in St. Bernard Parish with 850 square miles experiencing inundation in Plaquemies Parish.

The post-1947 improvements to the Jefferson Parish lakefront levee prevented serious damage there. Nonetheless, Flossy demonstrated that the lakefront, Industrial Canal, and lower river back levees were inadequate to prevent flooding from a strong hurricane and forcefully alerted the public to New Orleans's susceptibility to flooding. The city council promptly passed a resolution calling for an investigation of methods to improve the city's protection.[9]

Early in the 1957 hurricane season, Tropical Storm Audrey gathered its forces in the Gulf of Mexico and set out on a trajectory toward the Louisiana—Texas border area (see fig. 3.1). This powerful storm, although it did not strike the New Orleans region, offered a deadly warning to Louisiana officials and residents in other hurricane-susceptible locations. With winds of 125 miles per hour, Audrey roared ashore near the mouth of the Sabine River on June 27, 1957. Many residents of coastal Cameron Parish opted not to evacuate, and 556 perished under the pounding storm surge and waves that topped 20 feet at Cameron. The low-lying *cheniere* plains of southwest Louisiana offered residents little protection where surge reached 11.9 feet at Grand Cheniere and 10.9 at Pecan Island. The coastal communities suffered substantial damage, and Audrey delivered a powerful reminder of the serious threat that Louisiana faced.[10]

Both Flossy and Audrey pummeled the Gulf Coast shortly after a pair of tropical cyclones (Hazel and Carol) swept over the U.S. eastern seaboard in 1954. The two Atlantic coast storms had compelled Congress to authorize a survey of hurricane threats and damage to the eastern and southern seaboards. The two Gulf Coast storms accentuated the need to carry these studies forward. Between Flossy (1956) and Audrey (1957), the Senate Public Works Committee adopted a resolution to reassess the Corps' 1946 recommendations for hurricane protection along the south shore of Lake Pontchartrain in light of the most recent storm to hit the area—Flossy. In particular, the resolution called for developing plans for a comprehensive

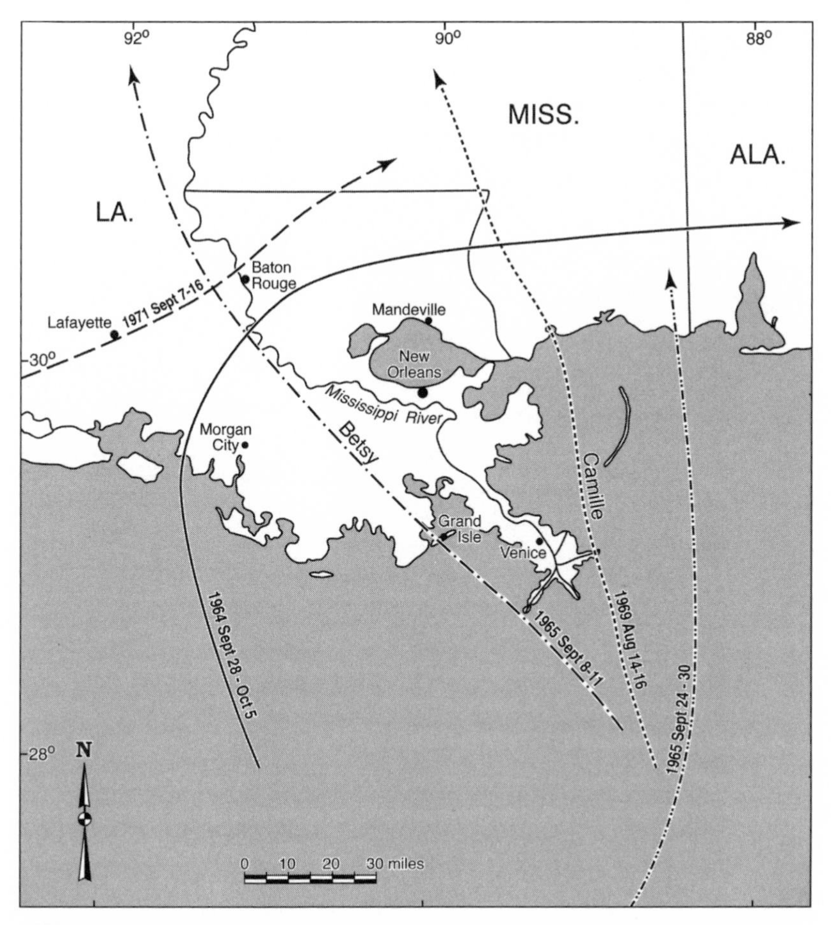

FIGURE 3.3 Hurricane paths, 1961–1971. After U.S. Army Corps of Engineers, 1972.

Lake Pontchartrain hurricane protection system. That task fell to the New Orleans District Corps of Engineers. In response to congress, the Corps issued an Interim Survey in 1962 that provided a context for subsequent planning. Yet, before Congress authorized funding for constructing any of the Corps' recommendations, two additional storms battered the New Orleans metropolitan area.[11]

The first of these storms, Hilda, arrived in October 1964. With peak offshore wind speeds of 130 miles per hour, the cyclone moved ashore over the central Louisiana coast overnight on October 3 with 100-to-120-

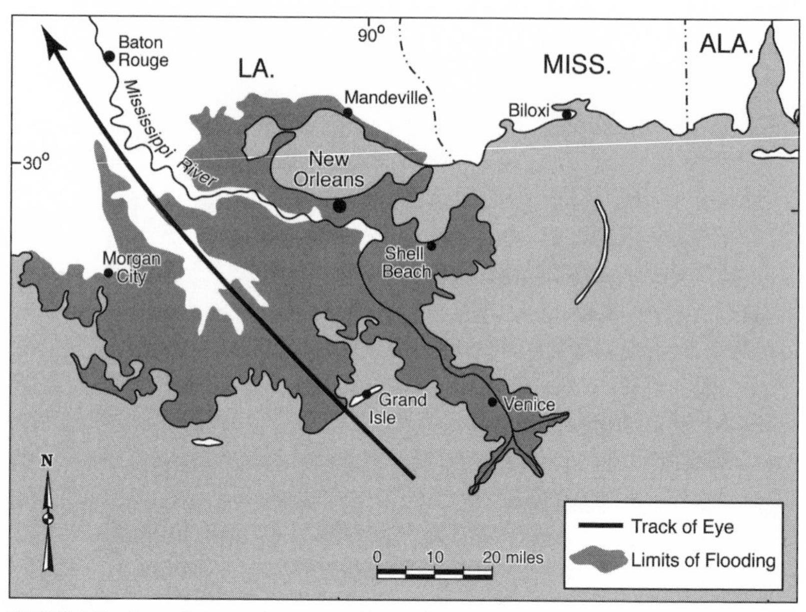

FIGURE 3.4 Hurricane Betsy track and area of inundation, September 1965. After U.S. Army Corps of Engineers, 1972.

mile-per-hour winds sweeping over inhabited areas (see fig. 3.3). It drove storm surges to ten feet or more and inundated more than three million acres of the low-lying coastal marshes. After moving over St. Mary and Iberia parishes, the storm veered eastward, passed over Baton Rouge in the early morning hours of October 4, and continued on an easterly track over Louisiana's Florida parishes (the parishes east of Baton Rouge) with winds below hurricane strength. The storm's leading edge produced a surge of six to seven feet along the north shore of Lake Pontchartrain and produced flooding in Madisonville and Mandeville. As the eye passed east into Mississippi, a strong cold front augmented the northerly winds of the back side and reversed the lake storm surge toward New Orleans. Wind-driven waves and surge caused substantial damage along the Orleans and Jefferson parish lakefront and along the Inner Harbor Navigation Canal. The weakened storm produced minimal damage to the frequently hard-hit delta parishes that lay well south of its route.[12]

Hurricane Betsy, which followed a year later, effectively reshaped public policy toward hurricanes in Louisiana and contributed greatly to

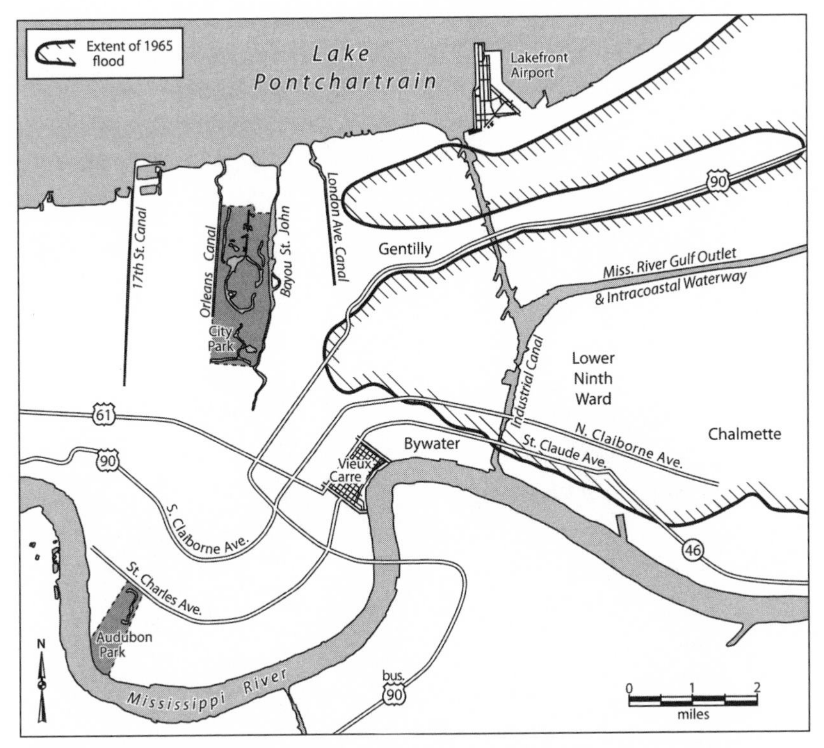

FIGURE 3.5 Area of flooding in New Orleans after Hurricane Betsy 1965. After U.S. Army Corps of Engineers, 1965.

national discussions about disaster response and a federal flood insurance program. After taking a lengthy track from the western Atlantic, across the southern tip of Florida, Betsy came ashore near Grand Isle on September 9 with maximum winds near 135 miles per hour (see fig. 3.4). Its eye followed a line just west of the Mississippi River and tracked west of Baton Rouge before turning north and losing strength. The storm surge and high waves destroyed most buildings on Grand Isle and topped eleven feet at the Mississippi River fishing community of Venice. Surge overtopped the back levees on the East Bank of the river and caused serious flooding in Plaquemines and St. Bernard parishes. Wind and waves lifted homes off their piers and piled them up along the levees that lined the lower river. Water overtopped and breached levees along the Inner Harbor Navigation Canal (Industrial Canal) and the Chalmette back levee. Drainage canals emptying

FIGURE 3.6 Flooding in the Ninth Ward, September 1965. Courtesy National Oceanographic and Atmospheric Administration.

into the Intracoastal Waterway also permitted surge to flow into the city. Consequently, there was extensive flooding in Chalmette, the Lower Ninth Ward, the Bywater, and Gentilly (see fig. 3.5). Some limited flooding along the lakefront occurred, but a back levee minimized extensive inundation in the city itself. Pumping of the flooded areas continued for several days. High water damaged nearly 10,000 homes in New Orleans and another 6,400 in St. Bernard and Plaquemines parishes. Surge devastated Grand Isle's more than 1,000 homes and camps and left 750 permanent residents homeless.[13]

Despite an evacuation of an estimated 500,000 people in the metropolitan region, the storm still caught many people in their homes. Reports of individuals awakened by water rising into their beds were not uncommon. Trapped by the rising water in low-lying neighborhoods, and with no lights to see their way to safety, many climbed into attics to await the storm's passing. Rescue efforts lifted some 11,000 people from the affected neighborhoods (see fig. 3.6). Within two weeks, pumps had largely expelled the floodwaters and only 2,300 evacuees remained at the main shelter in Algiers. Public utility crews and transportation companies had restored services within weeks. A month after the storm, the last of the Catholic schools reopened—a measure of the city's ability to return to near normal conditions quickly, even with a population of well over 600,000.[14]

President Lyndon Johnson promptly visited the scene and authorized two million dollars in relief funds.[15] Upon departing Louisiana, he proclaimed, "We want to cut all the redtape; we want to do everything to help the people of this great State get back on their feet," and he left four staff members in Louisiana to oversee the efficient delivery of federal aid.[16] Subsequent congressional hearings broadened the scope of concern from emergency aid and relief to include a federally operated flood insurance program.[17] Also, within days following the storm, Congress voted to fund hurricane protection plans submitted by the Corps earlier that spring.[18] Hurricane protection had captured the attention of public officials at all levels as well as the citizens of south Louisiana.

Although not terribly destructive to New Orleans, Hurricane Camille in 1969 delivered the most devastating impact to the Mississippi shore up to that time. It moved toward the coast with winds near 150 miles per hour, and the eye passed near the mouth of the Mississippi River on August 17 with winds near 200 miles per hour (see fig. 3.7). As it traveled over the bird's foot delta, it produced extensive flooding in Plaquemines and St. Bernard parishes. Camille delivered its most deadly blow to the Mississippi Gulf Coast communities. Storm surges exceeded twenty feet in Gulfport, Pass Christian, and Bay St. Louis, with Biloxi experiencing a surge between fifteen and twenty feet. Camille's assault demolished much of the tourist infrastructure along the beaches, dismantled port facilities, and damaged or destroyed tens of thousands of homes. Although an evacuation reduced the human toll, the Red Cross estimated more than 130 deaths. Robert Simpson, chief of the National Hurricane Center, proclaimed that, "by any yardstick, Camille was the greatest storm of any kind that has ever affected this nation."[19]

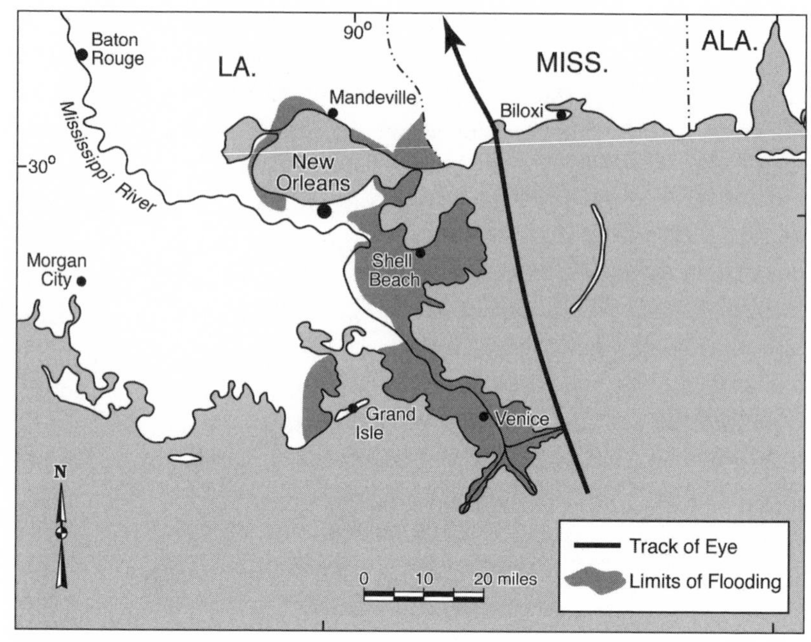

FIGURE 3.7 Hurricane Camille track and area of inundation, September 1969. After U.S. Army Corps of Engineers 1972.

The two late-1960s storms forced several major policy adjustments. The Corps of Engineers planners provided vital new storm information that could have been used to recalibrate their hurricane protection designs for the New Orleans District. However, it was several years before engineers factored in the meteorological data from these storms. Meanwhile, local officials vowed to take steps to prevent future storms from producing the devastation comparable to that delivered by Betsy and Camille. With public support running high, construction of hurricane protection seemed assured.

PUBLIC RESPONSE

On September 25, 1965, the House of Representatives Committee on Public Works convened hearings in New Orleans and Baton Rouge on the impacts of Hurricane Betsy. Since the storm had impacted southern Florida,

Louisiana, and Mississippi, there was a sufficient political base to prompt a federal inquiry. Florida reported damages in excess of $139 million, while Mississippi submitted an estimate of $19 million in losses.[20] Taking a more direct hit on a highly urbanized section of the state, Louisiana damage estimates topped $1 billion.[21] Herman Glazier, the executive assistant to the Mississippi governor, used the opportunity to revisit the Corps' feasibility study for Gulf Coast hurricane protection. He pointed out the substantial federal investment in ports and shipbuilding on the coast and the importance of the gulf fisheries and oil industry. Although previous investigations suggested that the two major storms of 1915 and 1947 did not justify regional hurricane protection, subsequent development substantially altered the equation. Furthermore, the back-to-back hurricanes of Hilda and Betsy, he argued, provided sufficient rationale to rethink the frequency of tropical cyclones striking the Louisiana-Mississippi coast. In particular, he urged Congress to develop a flood insurance program and to make its coverage retroactive.[22]

Governor John McKeithen of Louisiana opened his testimony with a reminder of the rich mineral wealth that lay beneath the state's coastal territory. Hoping to underscore the value of Louisiana to the country's energy demands, he claimed that the coastal region was perhaps "the wealthiest piece of land in the world." He then thanked Congress and the Corps for their commitment to building a substantial and effective flood protection system to protect the lower Mississippi from river flooding as a way of reminding them they were already engaged in flood defense efforts. And he acknowledged that Congress had helped fund a hurricane defense against Lake Pontchartrain waters. Despite these two massive efforts, he lamented that the most populous region of the state was not protected from surge driven on shore from the Gulf of Mexico. Like his counterpart in Mississippi, Governor McKeithen appealed for a federal flood/disaster insurance program to help citizens recover from events such as Betsy.[23] He pointed to the Federal Flood Insurance Act of 1956 and lamented that Congress passed the act but never implemented it. In his prepared statement, the governor noted that, if the Corps' "standard project hurricane" were to hit Louisiana, damages would exceed $175 million. He pleaded with Congress to fund the planning and construction of the New Orleans area hurricane-protection projects with all due haste.[24] The city's mayor, Victor Schiro, joined the governor in appealing for protection.[25] Testimony by the state public works director, Leon Gary, indicated that hurricane protection that would deliver benefits to Louisiana estimated at $475 million and would cost only $84 million. The benefits far outweighed the costs, he argued, and thereby fully justified the spending.[26]

Officials from St. Bernard, Jefferson, and other neighboring parishes offered their support for improved protection as well. Leander Perez, president of the Plaquemines Parish governing body, suggested that "every effort should be made by the U.S. Army Corps of Engineers and the parish of Plaquemines to expedite the construction of the proposed hurricane back levee" and investigations should be made into raising the river levees to offset surge that might travel up the river.[27] St. Bernard Parish officials pledged their 30 percent cost-share to the hurricane-protection project. Valentie Riess claimed that his parish considered its primary concern "that the Federal Government, in particular the U.S. Corps of Engineers give top priority and immediate action to the construction of hurricane protection levees along the Mississippi River-Gulf Outlet."[28] Jefferson Parish officials expressed appreciation for improvements already made to the lakefront levee system but claimed that the parish still faced a serious flooding hazard from the wetlands on the river's West Bank. For that section of the parish, they chimed in that they needed "more and better constructed levees."[29] Thus, there was solid support across the urban region for the federal government to proceed with improved protection.

The Corps of Engineers provided a view of the several phases of the hurricane protection system planned for coastal Louisiana. This system included the Lake Pontchartain and vicinity effort that had been improved following the 1947 hurricane with lakefront levees in Jefferson Parish. The one authorized project, the New Orleans to Venice section, called for a series of "back levees" that would loop around populated areas and connect with the river levees in Plaquemines and St. Bernard parishes. A Grand Isle project would protect inland areas along the lower Bayou Lafourche area—not Grand Isle itself. Finally, another project for Morgan City was on the drawing boards. Most of these projects were in the preliminary planning stages and had no appropriations for final design or construction at the time. To emphasize the importance of these projects to the region, the Corps' spokesman claimed that, if completed before Betsy arrived, they would have prevented much of the surge-caused flooding. A Corps map indicated that much of the populated sections of St. Bernard Parish, the lower Ninth Ward, the Gentilly and Bywater neighborhoods of New Orleans, and much of the lakefront areas of St. Charles Parish would not have been spared inundation.[30]

In subsequent hearings to consider a disaster relief bill, congressmen deliberated federal assistance in the form of temporary housing, small business loans, crop loss loans, and flood insurance. An underlying theme of these discussions was the fiscal challenges of disaster relief. Congressmen felt obligated to support disaster-relief bills for other sections of the country in order to be assured their region would get assistance in the time of need. Yet, the ad hoc nature of disaster relief produced inadequate responses. A cornerstone of an alternate plan was a federal flood insurance program. The idea behind this response was to assemble a federally subsidized reserve that could subsidize a type of insurance not available on the commercial market.[31] While Congress did not implement such a plan in the immediate aftermath of Hurricane Betsy, it passed the National Flood Insurance Act (PL 90-448) in 1968. This act called for systematic mapping of areas subject to an annual 1 percent flood risk (or a one-hundred-year flood), implementing local regulations to limit floodplain development (both riverine and coastal), and underwriting a federal flood insurance program available in communities that enrolled in the program. Core planks of this plan encouraged communities to minimize development in flood-prone areas and to shift the cost of recovery through insurance premiums to those who chose to live in flood-prone areas.

Despite years of pleading from local officials for such a program, participation at the local level moved very slowly. Only after the federal government undertook two important steps, and local governments put in place their portion of the program, were residents able to begin purchasing insurance. First, federal agencies mapped the one-hundred-year floodplains that defined the territory where flood insurance would be available. Second, Congress enacted legislation that required anyone using borrowed funds, with ties to the federal treasury, to purchase flood insurance. In addition, communities had to establish building codes for the floodplain in order for residents to purchase the federal insurance. Only after these steps were in place was insurance available.[32] Over time, south Louisiana became the region with the highest participation rates.[33]

Meanwhile, in 1965, Congress also approved funding for the Lake Pontchartrain Hurricane Protection system and the Grand Isle and Vicinity project. This enabled the Corps of Engineers to begin the detailed planning of a growing complex of hurricane-protection systems. These components received congressional authorization within months of Hurricane Betsy.

PLANNING FOR THE STANDARD PROJECT HURRICANE

As part of its post-1954 hurricane assessment, the Weather Bureau (which became the National Weather Service in 1970), in collaboration with the Army, developed what it called a "standard project hurricane." Based on frequency and magnitude of a tropical storm, the "standard project hurricane" was analogous to the Corps' "standard project storm" used to establish design standards for river flood protection. The "standard project storm" definition was "the most severe storm that is considered reasonably characteristic of the region in which the basin is located."[34] New Orleans fell within one of three zones of the Gulf of Mexico that the Weather Bureau assessed—a swath of the coast from near the mouth of the Calcasieu River to near the mouth of the Appalachicola River.

For New Orleans, the Corps of Engineers began its interim survey (1962) using a standard project hurricane as a storm with one-hundred-miles-per-hour winds and a return frequency of about two hundred years. Depending on its track, a storm of that magnitude could produce surges of over eleven feet along the south shore of Lake Pontchartrain and in the Chalmette area, over twelve feet on the north shore of the lake, along the lower delta parishes, and about thirteen feet at the Rigolets. For planning purposes, the Corps adopted a standard project hurricane tack that more or less followed Hurricane Betsy's path, moving from southeast to northwest with the eye passing just south of the city (see fig. 3.8). Passage of a storm of this magnitude, the Corps predicted, could inundate some 700,000 acres in the New Orleans vicinity, with approximately 460,000 of those acres in urban land uses. The Corps recognized that changing land use within the New Orleans metropolitan region and continuing urbanization and suburban sprawl would expose larger areas to standard project storm and thereby increase losses. Even though previous storms

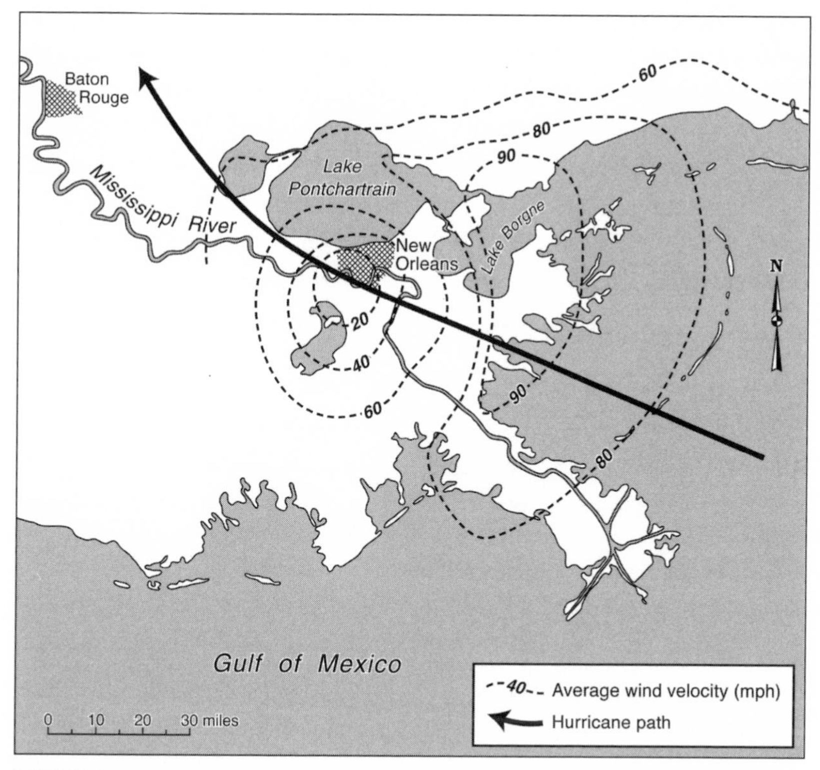

FIGURE 3.8 New Orleans area Standard Project Hurricane. After U.S. Army Corps of Engineers, 1966.

encountered minimal protective systems and caused considerable damage, planners forecast that a project storm could cause future catastrophic damages in as-yet-unbuilt sprawling suburbs in the neighborhood of $475 million.[35] Following hurricanes Betsy (1965) and Camille (1969), most components retained a standard project hurricane of one hundred miles per hour with a return frequency of two or three hundred years through the initial design stages. Designers applied different standards to different sections of the plan, based on local vulnerability and also land uses and populations protected by the levees. This created a piecemeal set of standards. Although the National Oceanographic and Atmospheric Administration (NOAA) adjusted its definition of a standard project hurricane for the New Orleans area that incorporated data from storms that occurred after Hurricane Betsy (1965) in 1979, the Corps retained the definitions developed in

1965 and 1966 for the Lake Pontchartrain and Vicinity and the New Orleans to Venice projects. The Corps finally incorporated NOAA's new definition into its plans in the 1980s.[36] While the Corps did adjust its calculations, it did not make those changes swiftly.

Beyond the standard project hurricane, several basic principles guided the initial planning effort. One fundamental concept was that the extensive, largely uninhabited wetlands to the east, south, and southwest of New Orleans had suffered "a minor degree of damage" from hurricane surge. Consequently, plans did not include these areas in the territories to be encircled by protective structures.

A second fundamental approach used by engineers was to employ protective structures around the more urbanized and built-up sections of New Orleans. Corps planners considered two options. The first they referred to as the "barrier plan." This option called for modestly increasing the height of existing levees and augmenting them with a set of barriers across the mouth of Lake Ponchartrain that would minimize the storm surge that could enter that waterbody. The second option was the "high-level" plan: it relied solely on raising the existing levees, along with construction of additional levees where they did not exist at the time, to protect against the surge anticipated in conjunction with a project storm. Initial estimates indicated that the high-level plan, with its larger demand for land and also a longer construction period owing to poor foundation materials for the levees, would be the more costly option. Consequently, the Corps' detailed study focused on the barrier plan.[37]

The barrier plan had several distinct components. The first was a new barrier levee extending across the peninsula that nearly enclosed the eastern end of Lake Pontchartrain. This levee would stand ten feet high and serve to block storm surge from entering the lake from the Mississippi Sound. A pair of barrier structures would close openings between the lake and the Gulf of Mexico at the Rigolets and the Chef Menteur Pass. These structures were to be movable gates connected to the barrier levee. They would remain open except when a hurricane threatened, and then operators would close them to block the surge from entering the lake. The concept behind this plan was that by minimizing surge, the Corps would have to spend far less on the levees surrounding the urbanized areas—in terms of both real estate acquisition and construction costs. In addition, by leaving the gates open except in the event of a storm, Corps officials concluded that the hurricane protection system would not affect normal salinity levels and biological conditions in the lake.[38]

Along the lakefront, the plan called for higher levees than the 1930s-era seawall to protect the growing urban area. The initial plan called for 5.5 miles of levee along the lakefront between Jefferson Parish and the Bonnet Carré spillway, standing ten feet high. This would protect a largely undeveloped wetland tract. Improvements to the Jefferson Parish levee since the 1947 hurricane were adequate, with a bit of riprap armoring. In Orleans Parish, plans called for raising the back levee along the lakefront to 11.5 feet and constructing a sheet pile flood wall along the Industrial Canal to a height of thirteen feet. The Corps' initial plan called for levees eleven feet high continuing along the lakefront from the Industrial Canal to the new barrier levee to protect the Citrus area. This levee would wrap back around toward the city along the Intracoastal Waterway at an elevation of thirteen feet. A sheet-pile flood wall along the Industrial Canal would enclose the western end. A ten-foot levee along the lakefront and a ten-foot levee along the Intracoastal Waterway would protect eastern New Orleans. St. Bernard Parish would receive protection in the form of a levee along the Mississippi River-Gulf Outlet (MRGO) with an elevation of thirteen feet.[39] The plan also called for additional strengthening of the seawall at Mandeville on the north shore.[40]

Given the inability of the marshy soils to bear a heavy load, levee construction would require multiple stages of levee building. The initial forecast projected six stages, or lifts. That is, crews would build the levees to the specified height, let them settle for two years, raise them to design height again, and then repeat until they stabilized at the desired grade.[41]

There was one additional key element in the barrier plan—the Seabrook Lock. This structure would connect Lake Pontchartrain and the Industrial Canal to protect property against high-velocity surge and also saltwater intrusion into Lake Pontchartrain. These two conditions were considered "adverse conditions resulting from the construction" of the MRGO. The Seabrook Lock would have dikes to an elevation of 13.2 feet. It would permit navigation into the MRGO and "prevent the entry of hurricane tides through the Mississippi River-Gulf Outlet."[42]

Initial cost estimates, based on 1961 dollars, indicated the Seabrook Lock would require over $109 million, the Lake Pontchartrain barrier plan an additional $64 million, and the Chalmette component $15 million. For the Lake Pontchartrain component alone, the Corps calculated annual benefits from flood damages prevented to exceed $47 million. Following Corps procedures, this ratio over the expected lifespan of the project justified construction, as did favorable ratios for the Seabrook and Chalmette components.[43]

The 1962 interim study recommended moving forward with the barrier plan. It also called for local interests to bear 30 percent of the costs (estimated at $41.2 million), supply all lands and easements, and also maintain the projects. Furthermore, the recommendations called for local interests not to hold the U.S. government responsible for damages owing to any failure of the flood-protection works.[44] This interim study stirred local organizations and individuals, and they began to mobilize, long before Betsy, to push for adjustments in the plan to accommodate their particular interests. Judge Leander Perez, a powerful political figure in the lower delta parishes in the 1960s, requested assistance from his congressman to adjust the proposed hurricane levee system. At the suggestion of Congressman F. Edward Hebert, the St. Bernard Parish Police Jury, for example, requested that the Corps enlarge the territory encompassed by the levee system to safeguard more businesses and landowners.[45] This was one of many local efforts to try to use federal funds to protect the maximum areas possible and thereby enhance local development and the parishes' tax base. Such efforts also enlarged the price tag, particularly in marshy areas where levees defined potential development.

In June 1965, Stephen Ailes, secretary of the Army, recommended the Corps' plan to the U.S. Congress, and on July 6, it came before to the Committee of Public Works. Thus on the eve of Hurricane Betsy, Congress had received a plan to protect New Orleans and vicinity from massive hurricane wave and surge.[46]

In the immediate wake of Hurricane Betsy on October 27, 1965, Congress authorized funding for the Corps to move forward with the plans recommended earlier that year.[47] As appropriations became available the following year, the New Orleans District directed substantial attention toward the barrier plan. Estimates for the Lake Pontchartrain portion of the project stood at more than $113 million from the federal treasury with an anticipated $52 million coming from nonfederal sources, $30 million of which would be a cash contribution from local interests. The initial work allowance appropriated in 1966 was $450,000, with funding rising to $1.6 million in 1967 and $4 million in 1968.[48]

Local and commercial interests quickly began to offer input into the planning process. An ad hoc citizens committee that supported the initial barrier plan proposal submitted comments to the Corps in November 1965. Following Hurricane Betsy, the committee argued that levee heights in eastern New Orleans were inadequate and recommended several height adjustments. It also recommended more floodgates than the Corps included in its barrier plan. Among the citizens committee's suggestions was a floodgate across the Mississippi River Gulf Outlet. The committee chair optimistically claimed that "the Army Engineers' plans, with our proposed revisions, in conjunction with Governor John McKeithen's plans for a levee across the Gulf Coast line of Louisiana should forever eliminate any danger of hurricane flooding to the populated areas of Louisiana."[49]

The Corps held a public hearing in St. Bernard Parish in December 1965 to allow local voices to be heard. Officials pointed out the frequency with which their parish suffered hurricane damages. Five storms had caused severe flooding since 1900. With only an eight-foot back levee, the parish defenses remained highly susceptible to overtopping. The parish sought inclusion of the levee along the Mississippi River Gulf Outlet as part of the hurricane-protection plan.[50] Parish residents voiced some disagreement not so much over whether or not there should be levees as over what areas required protection. Some sought to include far-flung communities like Yscloskey within the perimeter of levees. One engineer pointed out that, according to his calculations, the New Orleans area endured two hurricanes every three years. He also estimated it would take nine to twelve years to build the proposed sixteen-foot levees. Given the risk of eight hurricanes during that span, he urged the Corps to prioritize its resources and funds to protect the urbanized portion of the parish around Chalmette. Constructing levees around the more remote enclaves would increase risk to the "heart and head of the St. Bernard Parish," he claimed.[51]

Several local interests took prompt action and made commitments for their respective contributions. The Orleans Levee District, the St. Bernard Police Jury, and the Lake Borgne Levee District authorized "Acts of Assurance" in 1966. These Acts of Assurance committed the local interests to a

30 percent share of the first costs, provision of rights of way, infrastructure realignment to accommodate hurricane protection system, and maintenance after completion. Louisiana designated the Department of Public Works (today the Department of Transportation and Development) as the agency to coordinate local efforts.[52]

Work began promptly, and by September 1967, the local interests had completed interim floodwalls along the Inner Harbor Navigation Canal between Lake Pontchartrain and U.S. Highway 90. Additionally, levee construction proceeded, and crews completed a portion of the first lift of the Chalmette and Citrus sections that year. Officially, the work was 4 percent complete by May 1968.[53] By 1970, appropriations had reached $38 million and work proceeded in the Chalmette and eastern New Orleans districts, including the Citrus component.[54]

Meanwhile, early planning for the Grand Isle component moved forward. Initial appropriations of $1.1 million came in 1967, and another $816,000 in 1968. This enabled the Corps to secure Acts of Assurance with local interests there (February 1967) and hire an engineering firm to carry out investigations. By 1970, the New Orleans district reported it had completed 85 percent of the initial design memorandum for the loop levee along the lower Bayou Lafourche.[55]

EMERGING LOCAL CONTROVERSIES

Planning and initial construction of the Lake Pontchartrain and Vicinity Hurricane Protection Project took place in the era before the passage of the National Environmental Protection Act (NEPA) of 1969. This federal legislation required projects utilizing federal funds to prepare environmental impact statements that would identify any potential negative impacts of the projects and to consider alternatives that might produce less harmful consequences. Timing of initial construction projects and the passage of NEPA enabled the Corps to push forward several sections without preparing environmental impact statements. Projects remained largely engineering efforts, and the environmental considerations centered on concerns such as the load-bearing capacity of local soils, drainage characteristics, suitability of local materials for use in levees, and levee heights based on projected storm surge. Preliminary investigations sought to obtain information based primarily on civil engineering practice, not environmental protection.

Nonetheless, the Corps consulted with the Fish and Wildlife Service (F&WS) to consider the impacts of the Lake Pontchartrain barriers on wildlife. The wildlife agency's studies indicated that, with proper design, the Corps could build barriers that would produce negligible impact on marine life in Lake Pontchartrain.[56] Engineers completed numerous design memoranda before the end of 1970, and thereby relied on the joint Corps and F&WS investigations without facing the environmental impact requirements.[57] This all changed beginning in 1970.

In the immediate aftermath of Hurricane Betsy, local interests were very cooperative and moved quickly to approve agreements with the Corps. While often seeking enlarged protection for their respective districts, local organizations did not oppose improved protection. The Orleans Levee District even took the lead role in constructing much of the sheet pile levee along the Industrial Canal as part of its local contribution.[58]

Yet, conflicts emerged as well. Following the severe flooding caused by Hurricane Betsy near the Industrial Canal in 1965, a group of citizens filed suit against the U.S. government alleging that residents near the canal suffered property damage stemming from "negligent construction" of the Mississippi River-Gulf Outlet. Although the Corps secured agreements with local governments against flood liabilities for its flood-control projects, the lower court heard this suit since the MRGO was a navigation structure.[59] Even though the Corps expressed concern with a hurricane surge passing up the MRGO in 1963, its engineers argued before the court that MRGO was not responsible for the surge.[60] After several years of litigation, the U.S. District Court for the Eastern District of Louisiana ruled that the MRGO "did not cause in any manner, degree, or way induce, cause or occasion flooding in the Chalmette area."[61]

Despite a ruling against Louisiana plaintiffs in the MRGO case, there was a fundamental concern with the channel's impact on hurricane induced flooding. It took expression in local opposition to a new canal and lock proposed to pass through St. Bernard Parish. Numerous local interests remained convinced that MRGO contributed to local flooding, and consequently the federal government should absorb all hurricane protection expenses in the MRGO budget and absolve the parish of its cost share. St. Bernard Parish had quickly signed on to the cost-share arrangement after Hurricane Betsy, but as the MRGO controversy gained momentum, local interests sought to defer hurricane-protection costs to the federal treasury. Federal law did not allow the Corps to include levee costs in its navigation project, nor could it drop the local cost share already authorized by Congress and agreed to by local interests. Ultimately, St. Bernard residents allied with environmental groups to oppose the new canal and lock. This controversy commenced a series of delays in the canal project.[62]

Local environmental groups also began to view with skepticism the barrier plan and the section of levee connecting Jefferson Parish to the Bonnet Carré floodway. Over the next few years, and armed with new environmental mandates, local groups formed sometimes unexpected alliances to delay the hurricane-protection projects and in some cases force adjustments to the Corps' initial plans. The next chapter will consider these adjustments that occurred after 1970.

■ ■ ■ ■

LAKE PONTCHARTRAIN AND VICINITY HURRICANE PROTECTION, 1970–1990

As soon as Congress authorized the Lake Pontchartrain and Vicinity Project in 1965, Corps engineers embarked on detailed planning for the various components (see fig. 4.1a) Working with intense public interest to secure adequate protection for the next hurricane, both the Corps and its local partners initiated construction as soon as feasible. Initial activities focused on the '"barrier plan," which included levees and structures to impede storm surge from entering Lake Pontchartrain along with levees. Passage of the National Environmental Protection Act in 1969 and enactment of the federal requirement for projects to prepare environmental impact statements provided an opportunity for various local groups to challenge the barrier plan. In the most sweeping reconfiguration of the planning process, a federal court forced the Corps either to abandon the barrier plan or undertake additional environmental analysis in the late 1970s. This prompted delays in planning as the Corps opted to move forward with its "high-level" plan. The 1980s was a period with few hurricanes that might have delayed progress or forced design changes. Nonetheless, it was a turbulent decade for the project, which was running behind schedule and over budget. This makes the 1990 date a suitable breaking point in the Lake Pontchartrain narrative. Additionally, key decisions made in and after 1990 impacted the system in ways that contributed to the 2005 flooding.

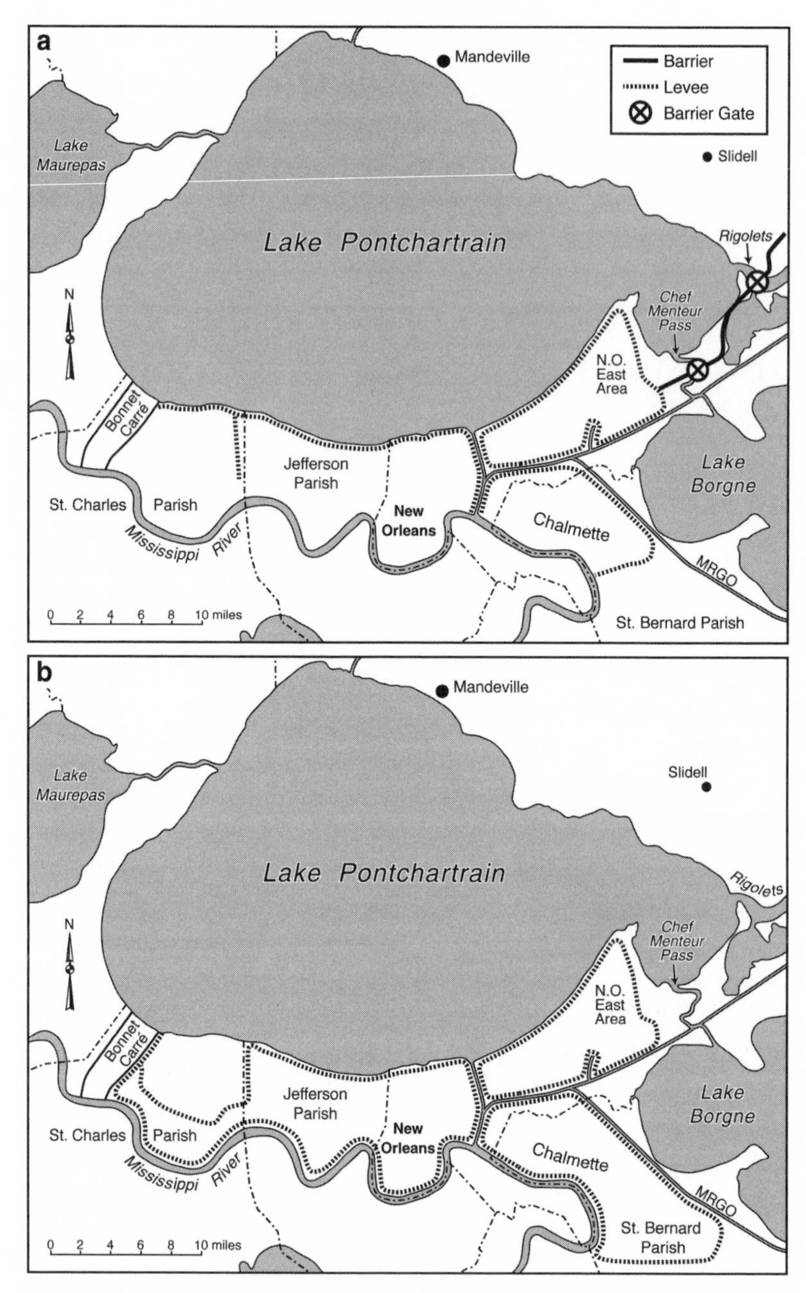

FIGURE 4.1 Hurricane-Protection Plans: (a) Barrier and (b) High-Level. After U.S. Army Corps of Engineers, 1966 and 1984.

DESIGNS AND INITIAL CONSTRUCTION THROUGH 1974

Inner Harbor Navigation Canal

Much of Hurricane Betsy's flooding resulted from the inadequacies of the levees along the Inner Harbor Navigation Canal (INHC or Industrial Canal). Corps engineers asserted that the levees along this canal were not designed to prevent a hurricane surge and that flooding from Betsy was not a surprise.[1] Following the deadly impacts of the 1965 flooding, this component became a top priority. One of the first design efforts focused on one site where levees failed in 1965. In March 1967, the Crops released its design memorandum supplement for the western levee along the Industrial Canal adjacent to the Bywater neighborhood that endured water depths of eight feet two years earlier. By that time, the Orleans Levee District had already begun work on fortifying the levees, and the Corps' design memo noted that "considerable work which will ultimately be incorporated into the overall project has already been accomplished by the sponsor."[2]

Drawing on studies by the U.S. Weather Bureau, the Corps adjusted its design standards from the pre-Betsy plans. Considering a storm with one-hundred-miles-per-hour (mph) winds, a forward motion of eleven mph, and a frequency of once in two hundred years, calculations predicted a surge of thirteen feet in the IHNC. Consequently, the Corps raised the levee heights from thirteen to fourteen feet to preserve a foot of freeboard.[3] Hurricane Katrina produced winds in excess of 100 miles per hour in southeast Louisiana as the storm made landfall and produced a peak surge at the IHNC of 14.3 feet.[4]

Intensive commercial land use along the Industrial Canal forced the adoption of sheet-pile floodwalls, rather than broad-based earthen levees. Designers called for the sheet pile to extend from 6.5 to 36 feet below grade.[5] Engineers noted inadequacies of this design with the modified project height and recommended an "I-type" floodwall where the height above ground is less than ten feet and a "t-type" floodwall where the sheet piling would stand ten feet or more above the ground. The I-type floodwall employs a single anchoring unit below the ground surface to add strength and prevent seepage, while the t-type uses a pair of anchors that flare away from the vertical floodwall underground to add greater strength. The design also called for a concrete cap for the steel sheet pile to reduce the corrosion anticipated from the saline water in the IHNC.[6]

The design memorandum also took into account local subsidence, noting that subsidence was a well-known, ongoing, and measurable process. For over half a century, various land-reclamation projects had encountered subsidence in the marshy soils of south Louisiana. Nineteenth-century schemes to convert wetlands into farmland ended in failure as the land subsided after drainage. The Corps acknowledged such experience: "large settlements of the ground surface have occurred in the marsh and swampland areas that have been reclaimed and drained, as a result of the shrinkage of the highly organic surface soils after drainage."[7] Engineers in the region also knew that highway right-of-ways built across the wetlands also subsided over time. Contemporary measurements placed subsidence at a rate of 0.39 feet per century. While most of the floodwalls adjacent to the Bywater were on more stable natural-levee soils, some sections had to deal with the prevalent subsidence issue. The designers concluded that "stability and settlement are major problems in the area."[8] Subsidence was an obvious and critical factor in design, and Corps engineers were aware of it and took it into consideration.

Work on the protection features along the Inner Harbor Navigation Canal proceeded according to the plans. Within three years, crews had completed much of the western floodwall and had initiated construction of additional floodwalls.[9] By 1973, 8.6 miles of concrete reinforced levee stood along the Industrial Canal and offered improved protection.[10]

Citrus—New Orleans East

Betsy also caused serious flooding in the lightly urbanized Citrus—New Orleans East areas, and this largely unoccupied area became the object of intensive planning as well. Anticipating that hurricane protection would enhance economic development, the Corps designed the levees for this component to withstand a surge produced by 100 mph winds. Hurricane Katrina at landfall produced winds above 100 mph and produced a surge of 15.5 feet where the MRGO and Gulf Intracoastal Waterway converge.[11]

The eastern lakefront area is susceptible to a double threat. Hurricanes passing to the east of the city can drive storm surge across the peninsula

TABLE 4.1 Adjusted Levee Heights

Levee Segment	Revised Grade—1967 (feet above mean sea level)	Original Grade—1962 (feet above mean sea level)
Citrus Back Levee		
West of Paris Road	14.0	13.0
East of Paris Road	18.0	16.0
New Orleans East Back Levee	17.5	16.0
Inner Harbor Navigation Canal Floodwall		
Seabrook to L&N RR	13.0–14.0	1.30
L&N RR to IHNC Lock	14.0	13.0
New Orleans East Lakefront Levee	12.0	10.0
Citrus Lakefront Levee	13.0	11.0
New Orleans Lakefront Levee	13.0	11.5
Jefferson Lakefront Levee	10.0	10.0
St. Charles Lakefront Levee	11.0	10.0
South Point to Intracoastal Waterway Levee	11.6	11.6
Barrier Embankment	9.0	9.0

Source: U.S. Army Corps of Engineers, New Orleans District, Design Memorandum No. 2. General Design, Citrus (New Orleans: U.S. Army Corps of Engineers, New Orleans District, 1967), 10.

separating the lake from the gulf. Additionally, wind-driven waves blown from the north can raise the water level, pounding the lakefront barriers. Designers had to take these dual factors into consideration. Forecasters calculated the Standard Project Hurricane would drive surge into the lake, raising its level six feet. The combined surge and waves would overtop all the pre-Betsy levees. Corps planners recognized the serious threat this posed to the areas of New Orleans that had subsided. Acknowledging that much of the city was at least two feet below sea level and portions as much as seven, the engineers determined that flooding up to sixteen feet in depth could occur with overtopping of lakefront levees.[12] To prevent this type of flooding, design heights of the Lake Pontchartrain plan varied from nine feet high for the barrier to eighteen feet for the New Orleans East Back Levee (see table 4.1).[13] Levee size depended on the potential wave height, which would be a function of fetch and wind direction.

Key to justifying the hurricane protection project was the economic benefit it would deliver. For the entire Lake Pontchartrain and Vicinity Project, the Corps calculated that 101,700 acres would receive protection. The combined benefit of damage prevention and future real estate development would total more than $65 million.[14] The Citrus component already had rudimentary levee protection but little development. Thus, it was one area where the future development side of the equation was particularly important. Its anticipated transformation from wetland to a suburban landscape was a critical component in the overall economic justification for the project.

For the eastern section, designed levee height was fourteen feet above mean sea level. Along this stretch, there was space to construct the stronger but subsidence-prone earthen levees.[15] The 1972 designs included additional adjustments from the preliminary plan authorized in 1965 and the 1967 calculations. Specifically, engineers raised the level of levee protection from ten feet to fourteen based on post-Betsy hydraulic studies. Additionally, the plan called for repositioning the levee landward of the Southern Railway embankment. This addressed several concerns: a shorter construction period so that protection would be in place sooner, preservation of forty-five lakefront camps (a concession to local interests), and less environmental disruption.[16] Although subsidence was a regional problem, particularly when constructing the heavier earthen levees, Corps designers concluded that although there would be subsidence, at least for the eastern segment of the New Orleans East levee conditions were "better than average."[17] In other words, subsidence would be less of a problem than in some other locations.

The Orleans Levee Board had constructed levees along one stretch of the New Orleans East component in 1956 and completed a lift in 1970, raising it to a height of 11.5 feet. The Corps planned to enlarge the levee by using sand dredged from the lakebed as a core. A semicompacted clay "blanket" would hold the sand in place and resist seepage.[18]

By 1974 crews were progressing in the New Orleans east segments of the plan. Work in 1974 saw 15 percent progress made on a 6.3 mile levee from Paris Road to South Point. Overall the Corps had completed 17.2 miles of first-lift levees in the New Orleans east area.[19]

Chalmette

Corps planners used the 100 mph standard project hurricane for the Chalmette component of the Lake Pontchartrain and Vicinity Project as well. Given the size of this project area, they used the two-hundred-year return interval. A storm with 100 mph winds could generate a combined surge and wave height ranging from about 11.5 feet to 13 feet on the gulf side of the New Orleans East, on the levees bordering Chalmette, and along the Industrial Canal.[20] Existing levees were only nine feet high and insufficient to deal with the design storm.[21] With these storm-driven wave heights, planners recommended levee heights ranging from 13 feet along the Inner Harbor Navigational Canal to 17.5 feet for New Orleans East and Chalmette.[22] These heights represented a net grade increase over the pre-Betsy plan (see table 4.1). These levees bore the brunt of Katrina's surge and waves.[23]

St. Bernard Parish, although large in total area, had a limited territory where development was viable (see fig. 4.1a). Land on the natural levee of the river is the highest surface and stands about ten feet above sea level. A relict natural levee along Bayou Terre-Aux-Beouf provides a second strip of land capable of supporting urban land uses. Much of the parish is about one foot above sea level and consists of marsh. A total of 30,000 acres lay within the planned hurricane-protection project, and of that only 4,500 were already developed.[24]

Given the topography and the amount of open space in the Chalmette component, the Corps plan called for a sizable water impoundment area. Anticipating as much as twelve inches of rainfall during a standard project hurricane (SPH), the plan called for 13,200 acres of land to collect this storm runoff to a depth of three feet. Following passage of the storm and the corresponding falling of the Mississippi River Gulf Outlet water level, drainage gates would permit outward flow into the neighboring canal. Thus, a sizable area had to remain undeveloped as a flood-retention basin. Local units bore the responsibility to acquire property and rights for this open space.[25]

The Corps argued that the combined levees and drainage structures would reduce the flood damages sustained by the parish from the average annual damage of $2.5 million to only $23,000. Coupled with future damages eliminated, the Corps' calculations suggested that "overall average annual flood damage prevented will amount to $7,255,000."[26] Such statistics were certainly appealing to communities and officials who had only recently endured Hurricane Betsy. Nonetheless, local interests, including St. Bernard Parish and the Lake Borgne Basin Levee District, requested a modification to the original Chalmette plan. These local organizations sought an enlargement of the area surrounded by levees.[27]

The initial hydraulic analysis for this enlarged levee-protected area, known as the Chalmette Extension, was completed by 1967. This component linked the Chalmette back levee along MRGO with the river levee below the community. The threat to Chalmette came from storm surge rising as much as eleven feet in the adjacent wetlands and from hurricane winds tossing the accumulated water over the existing structures, causing ponding in the developed sections of St. Bernard Parish.[28] The supplement would add approximately nineteen thousand acres to the initial plan.[29] The vast majority of benefits (74 percent) used "future development" and "land enhancement" to follow its construction as justification for this project.[30] This optimistic approach made the benefit-cost analysis highly favorable.

By 1974, crews completed the two control structures that would allow drainage from Bayou Bienvenue and Bayou Dupre to flow out of the Chalmette portion of the project. Overall, a total of 27.6 miles of first-lift levees and 11.9 miles of second-lift levees were ready for the next storm.[31] This component moved forward efficiently and without any environmental complications or controversies.

The Barrier Plan

Although never built, the barrier structures were prominent elements of the initial post-Betsy planning. When the Corps of Engineers selected the "barrier plan," their calculations demonstrated it was the less costly and also the more secure option. The barrier plan called for a levee spanning the peninsula that reached from New Orleans to the north shore and was to have included devices that could close the two water passages between the

Gulf of Mexico and Lake Pontchartrain. The design concept was to build structures that the Corps could close as a hurricane approached southeast Louisiana. By closing off the connecting waterways between the lake and the Gulf of Mexico, the structures would prevent the movement of massive amounts of storm surge into Lake Pontchartrain and thereby diminish the flood threat along the lakefront. By limiting the amount of surge in the lake, the Corps anticipated savings by reducing the height of levees to be built along the lakefront. A second option considered by the Corps was the "high-level" plan (to be discussed later) that did not have barriers at the two openings to the lake and compensated with higher levees along the lakefront. In 1965, the Corps found the barrier plan the better of the two options, but controversy eventually erupted over the impact it would have on the Lake Pontchartrain environment and forced a shift to the high-level plan in the 1980s.

RIGOLETS BARRIER · · · · The Rigolets was the northernmost of the two openings between the gulf and the lake. The movable barrier across the Rigolets, when closed, would link nine-foot levees that traversed the low marshy peninsula that enclosed the eastern end of Lake Pontchartrain with the New Orleans East system (see fig. 4.1a). A portion of the Rigolets levee system would use the preexisting U.S. Highway 90 embankment, where it achieved the design height. Entirely new levee structures would connect the raised highway embankment with the planned movable structures. Earth fill levees were possible in this section where space was not a limiting factor. Planners foresaw using sand dredged from the bed of Lake Pontchartrain as the levee core material and clays taken from the north shore as a protective cover. The control structure would have a gate at the Rigolets that could be lowered to close the channel when a hurricane approached. The gate, in conjunction with the levees, would preclude a storm surge from passing unimpeded into the lake and minimize the threat to the lakefront levees.[32]

As with the other components, the standard project storm for the Rigolets design called for a hurricane with winds of 100 mph moving at a speed of 11 mph. The levees across the eastern end of the lake could be subject to high water from either side, depending on the path of a storm. With the barrier closed, surge could overtop the levees and control structure and contribute to water-level rise on the opposite side. More important, water levels could rise in Lake Pontchartrain with the barriers closed owing to precipitation and runoff from the rivers feeding the lake's north shore. Following a storm, the difference between lake level and sea level could produce considerable current with the barriers' reopening. Design of the barrier structure had to take this hydraulic situation into account, and it called for stone riprap to line the closure channel to prevent scouring during regular tidal movement and to minimize erosion after a storm.[33] As in other areas of the Lake Pontchartrain and Vicinity Project, subsidence, poor soil quality, and settlement were a problem for the Rigolets components.[34]

The 1970 supplement provided some adjustments to earlier plans that provided some savings in terms of alignment and relocation of other structures, such as a segment of U.S. Highway 90. These savings provided additional justification, but the project did not receive a separate benefit-cost analysis. The authors of the 1970 supplement pointed out that the Rigolets barrier "is not a separable unit of the Lake Pontchartrain Barrier Plan."[35] This component did not provide direct protection for any urban area but offered a critical defense for the levees immediately adjacent to developed areas. The Corps opted to use the benefit-cost ratio of 12.4 to 1 calculated for the entire project to justify this component. Only planning and analysis of site conditions was under way on this component by 1975, well after the passage of NEPA.

CHEF MENTEUR BARRIER · · · · The Chef Menteur Pass is the other and southerly natural waterway linking Lake Pontchartrain and the Gulf of Mexico (see fig. 4.1a). It is a sinuous channel that transects the peninsula separating Lake Pontchartrian and Lake Borgne—a bay of the Gulf of Mexico. The Corps' barrier plan necessitated developing a means to close this channel to prevent excessive storm surge from entering the lake during a hurricane.

Using the standard project storm, the Corps' planners executed designs to contend with 100 mph winds and a storm moving through the area at 11 mph.[36] The plan called for building a control structure that would have a movable gate to be closed as a hurricane approached. Nine-foot-high levees would connect the control structure to the New Orleans East levee system and the raised embankment that carried U.S. Highway 90 across the peninsula and served the dual purpose of a levee for the Rigolets-Chef Menteur components.

Initial hydraulic analysis indicated that the control structure, when open, would not have an appreciable impact on the level of either Lake Borgne or Lake Pontchartrain.[37] Other design memorandum included brief mentions of environmental impacts, but the Corps did not include a similar discussion in its publication on the Chef Menteur component. The Design Memorandum included a letter from the U.S. Fish and Wildlife Service that concluded that the control structures at both the Rigolets and Chef Menteur would "have little appreciable effect on salinities in Lakes Maurepaus, Pontchartrain and Borgne. Therefore, no adverse effects on fish and wildlife resources in these areas are expected."[38] Prepared in the late 1960s without the type of field investigation that was becoming standard procedure for environmental impact statements, this type documentation had become obsolete by the time it was put to use in the 1970s. And environmental issues were to become a huge issue in subsequent years.

Included in the Design Memorandum is correspondence from the Orleans Levee District that indicates sharp opposition to one option to the barrier plan. The Corps considered a "floating gate" on the Mississippi River Gulf Outlet as part of "Alternate C" to this component. Local interests argued that this floating gate, when used as a hurricane approached, would interrupt traffic on the MRGO for several days and also burden them with onerous additional costs.[39] Navigation and other local interests presented powerful opposition to this option, which never materialized.

St. Charles Parish

The lakefront wetland area between the western boundary of Jefferson Parish and the Bonnet Carré spillway was largely undeveloped when Hurricane Betsy came ashore in 1965. Optimistic developers foresaw continued urban expansion across this wetland track, which local politicians saw as an expanding tax base. The same highway embankment that had offered limited protection in the past to Jefferson Parish ran along the St. Charles lakefront. In addition, entrepreneurs who sought to develop the LaBranche wetland had contributed to an ineffective lakefront levee system in the late nineteenth century. Corps planners included this area of potential development within the initial Lake Pontchartrain and Vicinity Project (see fig. 4.1a).

The design parameters for St. Charles Parish employed the same 100 mph storm but, given the lower population density, used a lower-risk return frequency of three hundred years. These criteria led to the proposal to build 5.7 miles of earthen levee along the lakefront with a height ranging from 12 to 12.5 feet above sea level. This represented an upward shift from the heights specified in the authorizing document and reflected the adjustments made following Hurricane Betsy to protect from high water of 10.5 feet along the levee.[40] The 1969 plan also allowed for the closure of a preexisting canal, which was deemed no longer a navigable waterway and thus eliminated the need for levees along the canal. This produced considerable design savings.[41]

The Fish and Wildlife Service and the Federal Water Pollution Control Administration had issued opinions that the overall project would not seriously impact wildlife or water quality in 1968. The Fish and Wildlife Service limited its comments to the impacts of the control structures at the Rigolets and Chef Menteur, not to the wetlands in St. Charles Parish. Thus, there was no specific consideration for environmental impacts in St. Charles Parish. In subsequent years, as the Corps reconsidered environmental impacts, wetland preservation became the key issue in this section of the project. The Corps' design memorandum indicated that "construction of the protective works covered herein will alter the existing terrain only to the extent of superimposing a hurricane protection levee with required contiguous features." In effect, they gave little consideration to the impact of the levee on the wetland behind the levee.[42]

Since the St. Charles component was part of the larger Lake Pontchartrain and Vicinity Project, the Corps did not conduct separate benefit-cost analysis. By 1974, the Corps had not initiated construction of this component but had prepared designs and completed initial site analysis.

PUBLIC INPUT AND ADJUSTMENTS TO THE PLANS

Public participation in developing the hurricane-protection plan began at the earliest stages and responses varied. Local input ranged from requests to enlarge protected areas, to proposals to defer local costs to the Corps, to efforts to block parts of the plan. In some instances, the Corps responded by making considerable changes with only modest push from local groups. In other situations, major and protracted conflicts ensued and sometimes rose to major court challenges. Ultimately, the give and take with local organizations produced delays in completing the project as well as design alterations—and these contributed to the rising costs.

The Corps held a public hearing in Chalmette just after Hurricane Betsy, in December 1965, to solicit input to the plan it had submitted to Congress earlier that year. Local interests, including public officials, industry representatives, and state officials, argued that the preliminary plan would not adequately protect St. Bernard Parish, and they requested a loop levee to encompass a larger area of the parish. Still visible damage delivered by Hurricane Betsy help make a compelling argument for an enlarged levee system.[43]

As a compromise, the Corps recommended and incorporated a larger loop levee into its plan. It encircled the more densely populated sections of the parish. Meanwhile, it opted to exclude a segment of levee around the remote community of Delacroix. Rather than take on the challenge of extending the levee around a small linear community with marsh for a building foundation, the Corps noted that most structures in the Delacroix area were already built to ride out storms or were "expendable." In addition, these areas with few residents and minimal property development had an extremely low benefit-to-cost ratio. Ultimately, the planners recommended that individual property owners could continue building flood-proof structures higher than the anticipated storm surge for this area—without the protection of levees.[44]

Public opposition to portions of the plan also arose in the early stages. Neville Levy, president of Equitable Equipment in New Orleans, voiced sharp and repeated opposition to part of the plan in 1967. In particular, he objected to the Seabrook Lock structure, an element of the barrier plan, which would control the flow of saltwater into Lake Pontchartrain during a storm. Levy wrote to the Corps and the local congressional delegation charging that the lock would cause surge driven up the Mississippi River Gulf Outlet to back up and damage industrial properties along the Industrial Canal. In many respects, he was repeating concern about the role the

MRGO played in flooding during Hurricane Betsy (discussion of the court case follows) but also claiming that the Seabrook Lock would exacerbate the problem. He conceded that if the Seabrook Lock had to be built, then the Corps needed to build a lock on the MRGO to impede storm surge pushing up this waterway. He called for a public hearing on the issue, but the District Engineer deemed a hearing "inappropriate at the present time." Colonel Thomas J. Bowen, the district engineer, pointed out to Levy that the Seabrook Lock was a vital part of the plan and would control saltwater flow into Lake Pontchartrain. If not controlled, increased salt concentrations could damage aquatic life in the brackish lake. In this case, the Corps took an environmental position to counter a local complaint.[45]

Local opinions obviously were not always aligned. In contrast to Levy, the Orleans Levee District and shipping interests opposed the construction of a control structure in MRGO. Shippers felt obstruction of the waterway during construction would be an economic burden; and they were uneasy with the Corps controlling the timing of closing the gate. Ultimately, the Corps opted not to build the control structure on MRGO.[46]

Despite its consideration of a control structure in MRGO, the Corps maintained an adamant position that MRGO did not contribute to flooding. In a post-Betsy lawsuit, Corps employees convincingly testified that the waterway was not responsible for damages in Chalmette. The suit field by St. Bernard Parish residents represented another means of public input into the process. Plaintiffs sued for damages to the properties caused by flooding produced by Hurricane Betsy. They alleged that the Corps of Engineers was responsible owing to its negligence in the design and construction of MRGO. The Corps argued that MRGO actually reduced the impact of flooding in Chalmette. Its experts also claimed that surge entered the Industrial Canal over open marsh and the Gulf Intercoastal Waterway, not via the MRGO. The Corps contended that damages were not the result of the deepwater channel. The courts accepted this position in its finding of fact: "The MRGO did not in any manner, degree, or way induce, cause, or occasion flooding in the Chalmette area. All flooding was the result of natural causes working upon local waters which have before threatened and caused flooding in the area due to the inadequate non-federal local protective features." Colonel Bowen made essentially the same case in 1967, stating that "based on exhaustive investigations of tidal phenomena associated with hurricanes, including Hurricane 'Betsy,' it may be stated categorically that the effect of the MR-GO on surge heights in the IHNC is inconsequential."[47] Despite the court's finding, the public continues to consider MRGO as a major contributor to hurricane-related flooding. This belief, in many respects, undergirds public distrust of official positions on the hurricane-protection system.

Public input also came in the form of appeals for more expeditious progress on the vital hurricane-protection system. While Betsy was still very much on the minds of local residents, interruptions to Corps appropriations prompted shrill cries for renewed funding. The Orleans Levee Board sent a telegram to Lieutenant General William F. Cassidy at the Corps of Engineers Headquarters in Washington in December 1968. It acknowledged reports that funding for "all flood control contracts" was frozen until the next year. Milton Dupuy, Orleans Levee Board president, demanded an explanation. He pointed out that "the lives and property of every man, woman, and child in our city is affected by this freeze."[48] Daniel Hall, assistant director of Civil Works for the Mississippi Valley, replied for General Cassidy, indicating that the Corps was reviewing projects resulting from the budget limits imposed by Congress. He continued by reporting that existing contracts would continue, as long as funding allowed, and that the Corps would make every effort to let new contracts as soon as possible.[49] Disruptions owing to congressional appropriations were one of many influences on the project's progress, although the Corps' 1969 annual report made no reference to funding interruptions.[50]

Local input to the levee design forced modifications, which tended to extend the construction schedule. Following a public meeting in early 1968 to discuss the proposed method for raising the Jefferson Parish lakefront levee, residents voiced sharp opposition to using an earthen berm and fill to raise the grade of their neighborhood. They argued that sheet piling would minimize the levee's footprint while providing greater flood protection and, at the same time, not devalue their property.[51]

Sometimes local organizations resisted the plans of the Corps but found that the state could intercede on behalf of the federal authorities. The St. Tammany Parish Police Jury was reluctant to provide assurances to meet its fiscal obligation toward the hurricane-protection project. Governor John McKeithen stepped in and, under state law, issued assurances for the parish in June 1972.[52] Through his action, the governor obligated St. Tammany

Parish to its local cost share. Nonetheless, public opposition to referenda to underwrite the parish's share resulted in a work stoppage on the floodwall in Mandeville. The Corps reported that the inactive status there was "due to the lack of financial participation in the project by the St. Tammany local sponsors."[53]

Unquestionably the sharpest lines between local and Corps plans involved Lake Pontchartrain. Growing public interest in environmental issues during the early 1970s sparked greater attention to the plans for the disposal of dredged material in Lake Pontchartrain. Unfortunately for the project's efficient progress, planning for the Lake Pontchartrain and Vicinity Project overlapped two eras of environmental legislation and requirements. The initial planning took place in a regulatory context that did not mandate thorough analysis of a project's environmental impacts. Corps officials since the late 1950s had consulted with the U.S. Fish and Wildlife Service about impacts to wildlife as required by the Fish and Wildlife Coordination Act of 1958. In addition, for post-Betsy planning, they sought comment from the Federal Water Pollution Control Administration regarding water quality impacts. With the passage of the National Environmental Policy Act in 1969, however, they had to shift from consultation to a more formal procedure. Federally funded projects had to conduct detailed environmental impact assessments to determine any adverse impacts and to evaluate options to minimize environmental damage from the project. As this act took effect in the early 1970s, procedures were ill-defined. Ambiguity in the act's requirements and the additional steps it mandated fomented delays and legal challenges. As an organization with a long-standing expertise in engineering, but not environmental protection, the Corps became the target of many challenges.[54]

The Corps released its Environmental Impact Statement (EIS) for the Lake Pontchartrain and Vicinity Project in August 1974. It contained a total of sixty-seven possible impacts. In summary, the EIS noted that most environmental impacts would be minimal and short-lived. Of critical importance was the salinity regime of Lake Pontchartrain, and the Corps asserted that the Seabrook structure would maintain appropriate levels. Loss of wetlands along the lakeshore would reduce the lake's productivity by reducing the release of detritus. Turbidity caused by dredging in Lake Pontchartrain would temporarily disrupt aquatic life but would create productive environments in the long run. The Corps suggested that "normal metropolitan expansion" was already encroaching on wetlands and conceded that the flood-protection plan would merely "hasten urbanization and industrialization." Thus, its work would not alter the environmental impacts, even though accelerating the forecast changes. Overall, the EIS concluded that the adverse impacts would be the loss of slightly more than 5,000 acres of wetland due to construction.[55] Yet, the EIS did not discuss the impacts of suburban sprawl across lakefront wetlands. Without future development, project planners would not have been able to project a positive benefit-cost ratio. The accounting for wetland loss was typical of the time but reflects a serious inconsistency in the Corps' analysis.

Shortly after the release of the first EIS on the hurricane-protection project, the Corps held a public hearing regarding environmental issues. After deflecting requests for public hearings on the entire EIS, the Corps had to hold a hearing on the relatively minor issue of dredge-spoil disposal. The EIS included several items among the sixty-seven impacts that dealt with extraction of dredged material but none that focused on the impacts of dredge spoil. Indeed, this was not a principal concern with the Lake Pontchartrain and Vicinity Project. Dredged sediment was the raw material for building levees and had very specific and beneficial uses. Nonetheless, Section 404 of the Clean Water Act (1972) required a hearing on this activity. While somewhat tangential to the project's core mission and procedures, this event opened the door for more general comments on the Corps' barrier plan.[56] Opponents to the barrier plan obviously relished an opportunity to make their views known to the Corps in a public forum.

The hearing opened with statements by several Corps engineers who described the barrier plan and its environmental impacts. They assured the audience that the barrier plan was not only the safer of the two but that it would be far less expensive and disruptive to residents of the urban area.[57] Corps spokespeople also summarized the amount of dredging anticipated with the project. Tens of millions of cubic feet of dredged material were necessary to build the earthen levees.[58] Since the dredged material was destined for the levees, very little of it would end up in navigable waterways.

Arthur Theis spoke on behalf of the Louisiana Department of Public Works, which supported the Corps. He referred to the critical need for the prompt completion of the hurricane-protection project following hurricanes

Betsy and Camille. He indicated that since the EIS had been filed with the Council on Environmental Quality, the state felt that there was no need to comment further on environmental impacts. In response to the Section 404 water quality focus of the hearing, he stated, "the term 'spoil disposal' is inappropriate in regard to this project" and "the areas designated as spoil disposal are coincidental with and pertain directly to the embankment location." Considering the other end of the process, dredge sites, he stated that "dredging locations and effluent discharges, should be directed toward the establishment of locations which will provide the least disturbance to the existing ecological balance of this area."[59] The state and the Corps, he suggested, could work to accommodate the highly productive commercial fisheries of the lake.

Opposition to the plan came from several fronts, not just environmental organizations. Edward Scoggin, a former commercial fisherman from St. Tammany Parish, opened his testimony by suggesting that the Corps' presentation was prepared by "Madison Avenue" types and presented by young men from other parts of the country, implying that they had no understanding of local circumstances. He argued against "bottling" up Lake Pontchartrain and claimed that all the Hurricane Betsy floodwaters came from outside the planned barrier. He claimed that the levee improvements since 1965 were sufficient to deal with a major storm. Most important to his Northshore neighbors, he asserted that the barrier plan would impede the drainage of the Pontchartrain-Maurepas basin. In the event of a major storm passing east of the Rigolets-Chef Menteur barriers, Scoggin claimed that the barrier would deflect the surge north of the levees into Slidell. He also charged that the barriers would contribute to the accumulation of pollutants in the lake and damage the commercial fisheries and also clam dredging.[60]

Mayor Frank Cusimano of Slidell spoke next. He indicated his support for Scoggin and presented resolutions from the St. Tammany Parish Municipal Association and the Slidell City Council opposing the barrier plan. He argued that the barrier plan would not protect New Orleans from water in the lake prior to a surge. He added that the barrier would inhibit economic development in the Northshore communities by precluding offshore service industries that could not build oil rigs to be floated out to the gulf through the barrier openings (an optimistic vision that never materialized). Mayor Cusimano also reminded the gathering that the residents of St. Tammany had voted "no" on tax refenda to supply the parish's local cost share.[61] The Slidell Chamber of Commerce also argued against the plan. Dave Martin, the organization's president, pointed out that the chamber considered the barrier a danger and had passed resolutions opposing it. He suggested that St. Tammany residents were being asked to subsidize a system to protect New Orleans that would not benefit the Northshore parish.[62]

The Fish and Wildlife Service, which had not objected to early barrier plans, raised doubts about the knowledge of the full impact of the barrier structures to the lake's ecology. Its statement indicated that the service was "concerned that there is an insufficient amount of biological knowledge available to accurately predict the effect of the Barrier Structures on the Movement of Organisms into and out of the lake." The Fish and Wildlife Service also supported the deferment of the St. Charles Parish lakefront levee since the hurricane-protection project would directly modify the basins of two bayous recently included in the state's Natural and Scenic River system. Their designation as scenic waterways made alterations contrary to state law. Thus, the state legislature's action to declare Bayous LaBranche and Trepagnier as scenic rivers effectively blocked the St. Charles component. Reflecting a rising concern with wetland preservation, the Fish and Wildlife Service added a strong statement that they opposed "needless destruction of wetlands."[63]

David Levy, the outspoken critic of the Corps who hailed from Slidell, resumed his opposition to the barrier plan at the public forum and pointed out that he represented two businesses in St. Tammany Parish. He claimed, erroneously, that in 260 years, no hurricane had driven saltwater from the lake into New Orleans and caused damage. He also noted, correctly, that the local population had rejected taxes to pay for its portion of the plan repeatedly. He concluded that there was no need for additional levees.[64]

Environmental groups also had their say. While they indicated support for hurricane protection, they offered several critiques of the project. The Clio Sportsman's League of New Orleans directed its comments to the Corps' EIS. Glenn Mercadal, on the league's behalf, argued that the EIS did not consider the value of wetlands in its benefit-cost assessment. He pointed out that if the economic analysis attached a dollar value to the marshes and swamps to be lost to rapid urbanization (and not just a benefit to future development), a recalculation would reduce or reverse the Corp's favorable ratio. His organization's main objection was the encirclement of large areas of undeveloped wetlands by hurricane protection levees. This, he asserted, was "unnecessary private land enhancement at the expense of the public and the environment."[65] The Orleans Audubon Society offered a similar criticism that building levees around undeveloped wetlands was both an environmental loss and a subsidy to private developers. They also made the case that wetlands should serve as a buffer against storm surge and that by building levees around the expansive marsh, the city would lose its cushioning effect. Taking sides with St. Bernard residents, the Audubon spokesperson argued that the MRGO was responsible for the "salinity problem" that necessitated a lock at Seabrook and that the same waterway contributed to flooding during Betsy. In effect, he suggested that the Corps had created problems and should not be counted on to eliminate them with the current plan.[66]

William Fontenot spoke on behalf of the Sierra Club, one of several environmental groups present. He pointed out what environmental groups saw as the flawed logic of ascribing benefits to future development while ignoring the value or benefits of the wetlands without development. He also drove home the point that much of the 73,000 acres within the planned protection barrier averaged only 1.5 feet above sea level and was unsuitable for the type of development projected by the benefit-cost analysis. He argued that once leveed and drained, these low-lying areas would have the "greatest potential for flooding." By encouraging development of these areas, the plan would place more people and property in harm's way.[67] The arguments of the Audubon Society and Sierra Club are amazingly prescient in the post-Katrina era.

Edgar Veillon, for the Louisiana Wildlife Federation, took a page from flood-protection planner Gilbert White and argued that "the completion of said project might, in fact, lull city and parish residents into a sense of false security while they await the elusive standard 'Project Hurricane.'"[68] Another spokesman, Art Crowe, made much the same point and argued that the plan cultivated a false sense of security and would lure development in low areas that would be devastated by a major storm like Camille. He suggested federal funds should not be spent destroying wetlands and luring development in flood-prone areas. Rather, he argued, federal subsidies would be better spent encouraging development on the north shore where there was higher ground—thus saving the marsh and making safe home sites.[69] Michael Tritico, on behalf of a group of scientists known as the Marine Environmental Researchers, took the security argument one step farther. He acknowledged the slow subsidence of the New Orleans area and stated that ultimately residents would have to abandon the city. He argued that the Corps needed to tell New Orleans residents the hard truth—that they lived in a vulnerable area and should relocate to higher ground. Sounding like some post-Katrina critics, he suggested that "the people of New Orleans are intelligent and courageous enough to begin considering an orderly plan for phased relocation of their residences to higher ground." Relocation, he concluded, would have a far more favorable benefit-cost ratio than a salvage operation after a future, powerful storm.[70]

Earl Colomb of St. Bernard Parish revisited that parish's obsession with the MRGO-induced flooding. He argued that if barriers could suppress surge in Lake Pontchartrain, similar structures should be able to dampen surge along the MRGO, and he alluded to an option considered by the Corps to include MRGO in the barrier plan.[71] William Gilmore of St. Bernard Parish endorsed Mr. Colomb's position.[72]

Guy LeMieux, president of the Orleans Levee Board, offered a counterposition. As a trained civil engineer with years of experience in hurricane protection in the New Orleans area, he could not be tarnished with the "Madison Avenue" label. He argued that he found the barrier plan the superior option and added that innumerable civic associations in Orleans Parish supported it.[73] Homeowners from the lakeshore neighborhoods also strongly endorsed the plan and emphasized the project's importance to protecting their investments.[74]

Obviously public opinion was divided, and the Corps had considerable leverage with its ranks of engineers and 70 percent of the budget. The Corps issued its Statement of Findings in August 1975. It underscored its conclusions that Lake Pontchartrain did indeed pose a significant flood threat. The findings also reiterated its belief that the barrier structures would create minimal disruption to the salinity regime of Lake Pontchartrain and its aquatic life. It acknowledged that further environmental study was necessary. Yet, the combination of its environmental impact statement, the public hearing, and its response to the public input satisfied Corps officials that it had met its legal obligation to proceed.[75]

Nonetheless, public opposition persisted. Col. E. R. Heiberg III, the district engineer, worked to win over the opposition. He flew a group of project opponents to the Waterways Experiment Station in Vicksburg to see the models and meet with the hydrological engineers who had analyzed the potential impacts of the barrier on the lake.[76] His campaign fell on deaf ears, and a coalition of business and environmental groups challenged the barrier plan in court. This proved to be the most influential public input into the Corps' planning.

In 1975, opponents to the barrier plan, united under the name of "Save Our Wetlands," filed suit to block this action.[77] Often labeled as a group of environmentalists, the opposition was more broadly based, as Louisiana state representative Edward Scoggin tried to make clear to Early Rush, the New Orleans district commander, in 1977. Scoggin claimed that the citizens of St. Tammany solidly opposed the barrier plan, as evidenced by the three rejected referenda to secure parish funding. He cautioned the colonel to read these actions as opposition not to hurricane protection but merely to the barrier plan. At the root of St. Tammany residents' concern was apprehension that the barriers, while protecting New Orleans, would redirect storm waters into their community. The lawsuit arose from the larger constituency.

Citing an inadequate environmental impact statement, the plaintiffs secured an injunction against the Chef Menteur Pass, Rigolets, New Orleans East, and Chalmette components of the barrier plan in December 1977. Several months later, the judge modified the injunction to allow construction to proceed on all portions except the Rigolets and Chef Menteur Pass projects.[78] This effectively blocked the barrier plan by eliminating the two key structures that were to impede storm surge from moving into Lake Pontchartrain during a hurricane. In the wake of this legal action, the Corps initiated a reevaluation of the alternate plan, the high-level option (see fig. 4.1b).

Although construction on elements of the barrier levee system had been under way for nearly two decades and about 51 percent of the work was complete,[79] the reevaluation study shifted the balance toward the high-level plan. The Corps concluded that the high-level plan would cause the "least environmental damage,"[80] would offer the greater benefits in terms of storm protection, and by the mid-1980s would produce a more acceptable

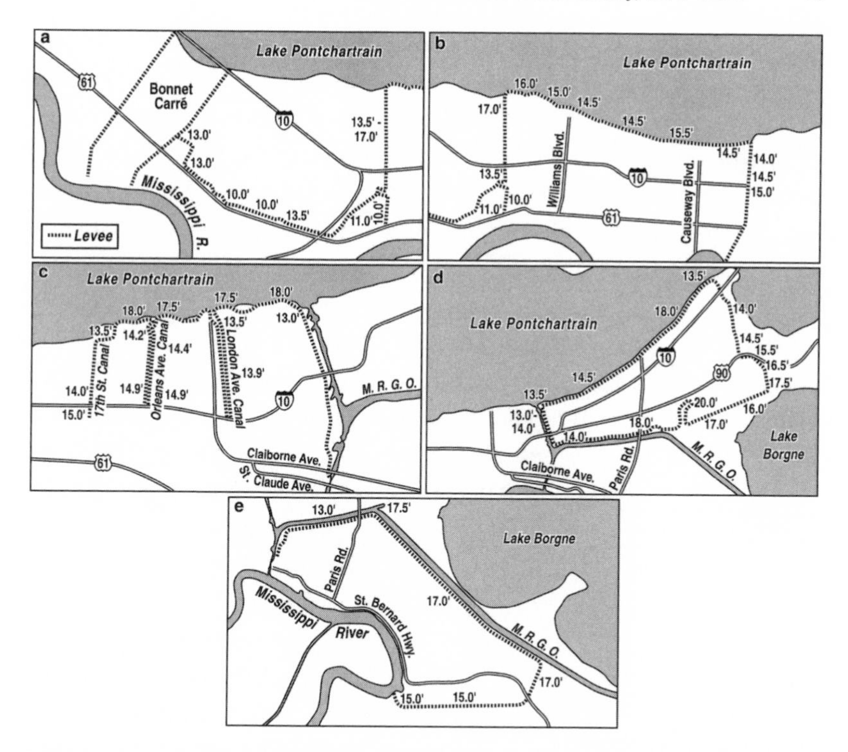

FIGURE 4.2 High-level levee system details and levee heights. (a) St. Charles Parish levees, (b) Jefferson Parish-East Bank levees, (c) Orleans Parish levees, (d) Eastern New Orleans levees, (e) St. Bernard Parish levees. Numbers indicate levee height above mean sea level. After U.S. Army Corps of Engineers, 2004.

benefit-cost ratio.[81] Given the local political climate, the high-level plan also was more acceptable to the public. Ultimately, this added delays to an already slow-moving project as the Corps revamped its detailed plans from a barrier to the high-level design. Louisiana congressman Robert Livingston appealed to the director of Civil Works to expedite the Corps' efforts. Acknowledging the necessity of revamping the designs, he wrote, "I wish to go on record as having protested these inordinate delays, and to urge you and the Corps to do everything humanly possible to speed up the intervals required for study, design, and construction of these most important projects."[82]

The revised plan incorporated several major adjustments. First, it formally dropped the already deferred lakefront levee in St. Charles Parish and recommended a levee set back from the lake and parallel to Highway 61 (see fig. 4.2a). This levee would be far less costly to build and would not disturb an immense wetland area adjacent to the lake. For Jefferson Parish, the high-level plan called for raising the existing levees to heights ranging from 13.5 to 14 feet (see fig. 4.2b) For Orleans Parish, the high-level plan required higher floodwalls along the lakefront—from 11 to 13.5 feet (see figs. 4.2c and 4.2d). In addition, this plan necessitated raising "the construction of measures to prevent overtopping of the outfall canals for the three pumping stations which are set back from the lakefront." Thus, the emphasis was put on preventing overtopping of all levees and not keeping storm surge out of the lake. Of course, this report indicated detailed designs were still necessary. Preliminary review indicated additional levee height ranging from 13.5 to 16.5 feet would be necessary in the New Orleans East/Citrus areas, as well as the St. Bernard Parish barriers (see fig. 4.2e). Multiple, and additional, lifts for the new levee heights would take additional time to complete.[83]

Following the court's order against proceeding with the barrier plan, the Corps had to revamp its plans. Engineers had to revisit, update, and prepare more detailed designs for the high-level plan that received official authorization in 1985. Design memoranda reached completion between 1984 and 1987. The revisions added considerable time to the construction schedule, and Congress eventually took note of the extended timeline.

DELAY OVERSIGHT

Slow progress on the Lake Pontchartrain and Vicinity Projects first attracted the attention of Congress in the mid-1970s. A decade after authorization, Congress requested that the comptroller general investigate the slow-moving project. Its spiraling price tag had risen from $85 million to $352 million, while delays had tacked on an additional thirteen years to the project's timeline. Even before the 1977 court order and its associated schedule and design revisions, work crept forward.

The Corps' initial projected completion date was 1978, but by 1976, it had revised that to 1993. In response to the comptroller general's investigation, the Corps reported several reasons for the delay.

1. Construction of levees, floodwalls, and roads took more time than anticipated. This resulted from poor foundation materials discovered "after project initiation." Instead of one year settling of new levees between lifts, experience demonstrated that 3 1/3 years was preferable. In part, adjustment of design heights after Hurricane Betsy mandated higher levees which contributed to the foundation problems and lags between lifts.
2. Local sponsors impeded acquisition of rights-of-way. The project called for local interests to supply rights-of-way, but sometimes the local levee boards and parish governments either refused to provide the proper property or prioritized project components at odds with Corps plans. Such delays, the Corps claimed, were beyond its control and had added years to the project.[84]

The report treated both the St. Charles and St. Tammany lakefront construction projects as delays. It noted that "citizen actions" had indefinitely deferred the seawall project in Mandeville. It also pointed out that environmentalists opposed the lakefront levee in St. Charles Parish and that the State of Louisiana complicated matters by designating two bayous as part of its Natural and Scenic River system.[85]

While not providing an in-depth assessment of the still pending *Save Our Wetlands v. Rush* litigation, the report acknowledged that "the litigation has delayed the timely completion of the project" and that it could halt the construction of the control structures.[86] Indeed, when the court issued its order blocking the Chef Menteur and Rigolets barrier components, the project fell further behind schedule and prompted another federal investigation. In 1982, the General Accounting Office (GAO) sought an explanation for the cost increases, construction delays, and the Corps' poor performance. By this time, the projected completion date had been set back to 2008, with about 50 percent completed at the time of the investigation, and the cost estimate soared to $924 million.[87]

Forced to reevaluate its choice between the barrier and high-level plans, the Corps reported to the GAO that it was reconsidering the second option. In response to the 1977 court order, the Corps halted all efforts directed toward the barrier plan, and this added to delays. Also, it was preparing a revised environmental impact assessment that prolonged the process. The Corps indicated to the GAO investigators that, for five years, it had been unable to prepare an EIS that was satisfactory to the court. Thus, it directly implicated the court in the delay. Engineers had initiated detailed design work for the high-level plan, and those were taking additional time. The high-level plan would necessitate additional property acquisition—a delay pointed out in the 1976 investigation—and additional construction time to allow for more lifts of the higher levees.[88]

Another foreseen delay involved construction on the outlet canals for Orleans Parish. The standard project hurricane indicated that the existing levees along the canals were too low to hold back storm surge. The Corps was engaged in discussions with local interests over options to deal with this problem. One option was to raise the levees along the canals to contend with the higher surge anticipated with a standard project hurricane. Another option was to build movable gates at the mouths of the canals that could be closed in time to block the surge at the lakefront. This latter option would require auxiliary pumps to lift water from the canals into the lake. Corps personnel held discussions with the local sponsors in 1980, but they could not agree on an option. This added uncertainty to the project's schedule.[89]

HURRICANE-PROTECTION SYSTEM PROGRESS

Despite delays encountered with some components, construction crews made considerable headway in other areas. Storm surge continued to pose a considerable threat, but levees constructed to the initial post-Betsy design heights rose around most areas that faced high risk. Portions of the New Orleans East/Citrus and Chalmette levees were in place by the time Camille struck the Mississippi coast in 1969 and fended off some damage. At the time of the congressionally mandated assessment in 1976, the Corps estimated the project was just over 23 percent complete. Completed work included 8.6 miles of reinforced concrete floodwalls and 17.2 miles of first-lift levees in the New Orleans East area, plus 1.2 miles of floodwall, 27.6 miles of first-lift levees, and 11.9 miles of second-lift levees, and two control structures in the Chalmette area. Given that the initial plan projected a 1978 completion date, the progress to that time was far short of the early objective. Consequently, New Orleans and vicinity remained susceptible to a hurricane strike.[90]

By 1982, Corps officials reported considerable progress, revealing to the GAO that it had completed 70 percent of the levees and floodwalls.[91] For the entire Lake Pontchartrain project, however, the Corps indicated in its annual report that work was only 51 percent complete. According to the Corps' 1983 status report, the Lake Pontchartrain project consisted of eighty miles of levees and floodwalls. Approximately sixty-eight miles were complete or had first- or second-lifts.[92] The entire St. Charles and Northshore components remained part of the authorized plan but were on indefinite hold at that time.

One bottleneck had been eliminated by 1990. Funding for the Mandeville project was authorized by the Water Resources and Development Act of 1986, and preliminary designs were under way. While one stalemate had been nudged into action, St. Charles Parish remained in an inactive status.[93] Nonetheless, the Corps was able to report continued progress. The 1989 annual report stated that the entire Lake Pontchartrain project was 77 percent complete.[94] With slightly more than three quarters of the revised plan in place, the Corps pressed ahead, although the project had acquired an air of a perpetual process with no real end in sight.

In the late 1980s, the Corps and local sponsors had yet to agree on decisions about how to handle the outfall canals from the New Orleans municipal drainage system. This left a sizable weak link in the system. Engineers and local sponsors considered two options: floodgates at the mouth of the canals that could be closed during a hurricane and raised levees. Engineers from the New Orleans District and staff at the Waterways Experiment Station (WES) in Vicksburg favored gates at the mouths of the canals and seriously questioned the viability of parallel levees along the outfall canals. They deliberated the challenges to the gate designs and conducted experiments with hydraulic models to select a viable butterfly gate design.[95]

With the mouths of the canals closed, particularly during a hurricane accompanied by heavy precipitation, it would be impossible for the local drainage authority to expel runoff into the lake. By closing the gates, the canals would become impoundments and ultimately overflow back into the city—completely reversing their purpose. The Sewerage and Water Board of New Orleans expressed vigorous opposition to any plan that would impose a "pre-arranged set of conditions" for closing gates at the mouths of its drainage canals. Selection of a plan for the outfall canals, therefore, had to strike a balance between adequate flood protection and "efficient" operation of the Sewerage and Water Boards pumping capabilities. By the late 1980s, the high level plan and its parallel levees along the outfall canals had become the economic choice as well.[96] The Corps' 1990 design memorandum for the 17th Street Canal accepted the Parallel Protection plan. This option called for raised floodwalls along the outfall canal. Given space constraints, a concrete floodwall atop existing levees was necessary rather than the bulkier earthen levees.[97] Despite misgivings by the New Orleans District engineers, similar parallel levees became the choice for the other outfall canals as the high-level planning moved forward.

STORMS DURING CONSTRUCTION

In the years between Betsy and 1990, New Orleans avoided devastating hurricanes, although the region did endure several tropical storms. Hurricane Camille in 1969 earned the reputation as the most devastating storm to strike the Gulf Coast. A small but powerful category 5 storm, it roared ashore east of New Orleans, obliterating much of the Mississippi shore.[98] The potent storm's eye passed east of Louisiana and drove a massive surge that flooded the communities along the southern Mississippi coast. While Mississippi endured extraordinary damage, there was considerable flooding within the New Orleans District. Surge washed over 900,000 acres in southeast Louisiana. Winds in excess of 200 mph pushed surge and waves over the protective levees that surrounded the higher natural levee in Plaquemines Parish. Water rose as high as sixteen feet deep and inundated approximately 66 percent of the parish.[99] St. Bernard Parish also sustained major damages, but post-Betsy improvements limited the most severe damages to areas outside the protective levee system. Storm surge covered about 94 percent of the parish's marshlands. Of the over 251,000 acres under water, urban and agricultural uses claimed only about 1,000 acres. A storm surge eight feet deep covered the fishing community of Yscloskey. Most homes in the marshlands stood on piers and withstood the storm, although some houses had water four feet deep in their lower floor. The relatively small population, along with floodproof construction in the fishing communities in the marsh, minimized the storm's impact there.[100] Flooding inundated about 51,000 acres in Jefferson and Orleans parishes combined, and flooding produced serious impacts in the Bywater neighborhood, a highly urbanized district of New Orleans near where the Inner Harbor Navigation Canal crosses St. Claude Avenue—and an area also flooded by Betsy. There was also extensive flooding in the Citrus/New Orleans East areas. Levee construction was under way in this area but had not progressed sufficiently to encourage major residential development. The lakefront levees held off surge in Orleans and Jefferson parishes.[101] St. Tammany Parish sustained some damage, particularly in low-lying areas near the Rigolets, but the path of the eye east of the parish pushed waters away from the north shore and limited lake surge flooding.[102] This storm reinforced local concern about hurricanes while inflicting its worst damage on neighboring Mississippi. Modifications to the initial plans for Plaquemines and St. Bernard parishes followed the powerful impact of Hurricane Camille (see chapter 5).

The 1970s produced few major hurricane threats to the New Orleans area. In September 1974, Hurricane Carmen came ashore along the central Louisiana coast and veered to the northwest. It delivered heavy rains and blustery winds but was sufficiently distant that is was not a major factor to the state's largest urban area. Its most extensive damage befell the sugarcane crop. Hurricane Bob made landfall in July 1979. It moved ashore as a minimal hurricane west of Grand Isle and spared New Orleans major harm. Neither of these storms produced any serious flooding or disruptions to the Lake Pontchartrain project.[103]

A spate of hurricanes struck Louisiana in 1985 (see fig. 4.3). That August, Danny moved inland over Grand Isle, unleashed waves up to eight feet high, and caused serious erosion to the barrier island. New Orleans escaped serious damage. Elena followed a west-northwest track from near Mobile, Alabama, and moved on shore near Biloxi in early September. It raised Lake Pontchartrain levels as much as four feet. Winds declined from category 3 (110 mph) to tropical storm strength as it moved north of New Orleans. There was modest damage to the city.[104] Juan followed a loop-de-loop course over the Gulf of Mexico. One loop took it on shore along the central Louisiana coast in late October, and then it veered back over the gulf on a southeasterly course. The eye followed the Louisiana coast, moved over

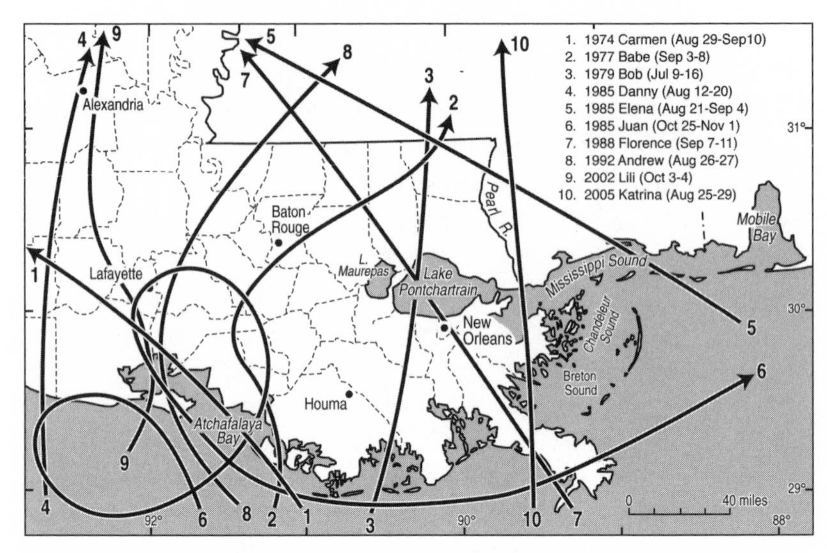

FIGURE 4.3 Hurricane paths, 1980–2005. After Yodis and Colten 2007.

the mouth of the Mississippi, and looped on shore near Mobile. Juan's winds and storm surge drove water up the canals and waterways of West Bank Jefferson Parish (see chapter 5). Incomplete locally built hurricane-protection works allowed flooding on the West Bank. Flooding also occurred on the Northshore of Lake Pontchartrain. Juan and its flooding, as much as any event to that point, broke stalemates with local interests and prompted the resumption of critical hurricane-protection works (see chapter 5).[105]

PROJECT FUNDING

In the immediate wake of Hurricane Betsy, there was broad-based support for funding the Lake Pontchartrain, Louisiana, and Vicinity Hurricane Protection Project, and appropriations began flowing in 1967. Initial budget justifications cited annual benefits of over $58 million dollars—for both flood damages offset and for potential development and a highly favorable benefit-cost ratio.[106] Through the federal fiscal year of 1969, Congress had made more than $12 million available. Local sponsors expressed serious concern about possible interruption of funding in late 1968. Milton Dupuy of the Orleans Levee Board sent a desperate telegram to the chief of engineers upon hearing the Corps had frozen all construction. Lieutenant Colonel Daniel Hall, of the Mississippi Valley Division, replied that the Revenue and Expenditure Act of 1968 had imposed expenditure limits and the Corps was assessing the situation nationwide. He indicated to Dupuy that the Corps likely would not be able to meet all current contracts and that new bids for new contracts were on hold.[107]

Despite concern by local sponsors and congressional budget control efforts, funding for the Lake Pontchartrain project continued to flow. Without exception, Corps officials noted that Congress made sufficient funds available, and they recalled no interruptions to work due to lack of appropriations.[108] Nonetheless, budget-conscious administrations could send fear through the ranks of New Orleans-area officials when they discussed controlling federal spending. In 1971, for example, the Nixon administration proposed only a small fraction of the funds necessary to complete the barrier plan by its original 1978 completion date. Denis Barry, of the Regional Planning Commission, wrote to President Nixon to remind him that the "dread of inundation from another cataclysm is very real" and to urge him to reconsider funding the project "so that our protection system for the lives of millions of people in this area will become a reality."[109] Federal budget challenges remained an issue in the timely construction of the hurricane-protection system.

Swelling project budgets became obvious early on in the process, contributing to federal concerns about the total cost. Rising costs were a focus of the comptroller general's investigation in 1976. It noted that the projected costs rose from $85 million in 1965 to $352 million in 1976. The Corps offered several explanations for the rapid escalation. A sizable component of the changes related to engineering adjustments. Some $54 million of the boost stemmed from increased levee size mandated by the upgraded standard project hurricane. New elements to the plan, such as the extended levee system around Chalmette, added another $13 million. By far, the largest cause of project cost inflation was national economic growth. As the economy grew, cost of materials, labor, and other components of the budget swelled. This added substantially to the project's expense, and with each delay, appropriations fell farther behind actual expenses. Of the total $267 million projected increase, nearly 68 percent or $183 million resulted from general economic growth.[110] The 1976 report also noted that additional costs remained to be factored in, such as the flood-control features for the Orleans Parish outfall canals. An area of grave concern to the comptroller general was the ability of local sponsors to pay their share of the spiraling costs. They had signed on to the project expecting to pay 30 percent of a projected budget of about $122 million. The projected budget's growth above $500 million imposed extraordinary burdens on the local sponsors. The comptroller general offered a blunt assessment: "It is questionable whether local jurisdictions will ever be able to pay their 30-percent share." If locals were unable to meet their obligations, that would mean additional federal costs to complete the project.[111]

The comptroller general's report stated quite clearly that funding had not contributed to project delays. It noted that the project maintained priority status within the New Orleans District office. In fact, funding outpaced expenditures. The report pointed out that, in several years, the Corps had been unable to spend all its hurricane-protection appropriations and rolled funds over to the next fiscal year.[112] When the General Accounting Office conducted its 1982 investigation, it made no comment on disruptions to the project schedule related to funding interruptions. As of 1982, Congress had appropriated $131 million for the project and local interests had anted up $40 million.[113]

According to the Corps' budget tracking, through 1972, a generous Congress enacted a budget allowance that exceeded the president's request all but one year (see table 4.2). That situation reversed in 1973, when for the first time Congress appropriated less than requested. Through 1990, the enacted work allowance fell consistently below the president's and the congressional committee's recommendations. During this time, the project budget was at its maximum levels, and the difference between requested funds and appropriations was only a few million dollars annually when budgets averaged more than $15 million per year. In 1974, despite a presidential request and congressional recommendation of $6.4 million, the Corps received no work allowance. In 1979, the project received a recommendation for zero funds from both the president and Congress. Appropriations between 1975 and 1989 ranged from $2 million to more than $36 million. Between 1982 and 1988, the work allowance hovered around $15 million annually.[114] Despite recollections by Corps personnel that there was no funding shortages, the New Orleans District received less than requested by the president most

TABLE 4.2 Lake Pontchartrain and Vicinity Budget, 1967–90

Fiscal Year	President's Request	Congressional Committee Recommendation	Work Allowance (Enacted)
1990	39,898	39,898	35,639
1989	40,400	40,400	36,384
1988	17,000	17,000	14,784
1987	16,000	16,000	15,375
1986	25,000	25,000	20,288
1985	17,500	17,500	15,100
1984	16,800	16,800	15,800
1983	18,000	18,800	14,800
1982	15,000	15,000	15,000
1981	10,800	10,800	9,600
1980	11,000	11,000	10,000
1979	0	0	0
1978	12,400	11,300	10,000
1977	12,000	12,000	10,700
1976	29,350	21,350	19,985
1975	3,300	3,300	2,100
1974	6,400	6,400	0
1973	20,000	na	17,500
1972	4,555	na	10,946
1971	8,250	na	11,250
1970	6,000	na	8,050
1969	7,800	na	6,274
1968	2,300	na	4,000
1967	450	na	1,600

Source: U.S. Army Corps of Engineers, "New Orleans District, New Orleans Projects—Actual Costs," unpublished spreadsheet, 2005.

years through the 1980s. Apparently, the shortfall was not sufficient to disrupt progress, as the earlier investigations suggested.

Despite a series of administrations committed to controlling the cost of government and spending for overseas military efforts during the Vietnam conflict, Congress remained committed to funding the Lake Pontchartrain area hurricane-protection project. Most budgets were slightly below the requested levels but never fell low enough to hamper progress.

CONCLUSIONS

Initial design efforts focused on the barrier plan, and early construction also worked toward that end. Public pressure against the barrier plan prompted considerable conflict and ultimately a suit charging inadequacies in the 1974 Environmental Impact Statement. When the federal court imposed its order blocking the barriers in 1978, the Corps had to reorient both planning and construction efforts toward a substantially different goal. Huge delays ensued. The New Orleans District had to issue a new environmental impact statement, prepare new designs, and work with local sponsors to secure their continued commitment to the reworked plan. Although funding remained adequate and construction progressed, as of 1990 the project was only 77 percent complete.

■ ■ ■ ■ ■

PROTECTING THE DELTA, THE WEST BANK, AND THE COAST

The communities along the narrow band of natural levee that follows the lower Mississippi River endured devastating impacts from Hurricane Betsy in 1965. Exposed to surges blown across the spindly finger of land that extends into the Gulf, the lower delta has been one of the most susceptible and most frequently impacted zones in Louisiana. For that exposed location, Corps engineers had to design and build "back levees" that would loop around the slightly higher ground and connect to the existing Mississippi River levees. On the West Bank of the river across from New Orleans, different conditions prevailed. There, proximity to the ocean or Lake Pontchartrain was not the issue. Storm surge, nonetheless, remained a concern. Canals, lakes, and natural channels that linked the low-lying wetlands south and west of the Mississippi to the Gulf of Mexico made the West Bank urbanized area susceptible to hurricane-driven surge. In addition, at the southernmost extent of Jefferson Parish, the resort community of Grand Isle suffered all-too-frequent devastation from hurricanes. While the island was virtually impossible to protect with levees, Corps planners sought to minimize destruction to the residences on this eroding barrier island by other means. Populated areas up Bayou Lafourche from Grand Isle also faced threats and warranted a hurricane-protection project. This chapter will examine the four components that addressed these distinct locations and will reveal tremendous variation in terms of local cooperation and timely progress in completing these components of the regional plan.

ADDING PROTECTION TO THE LOWER DELTA

The preliminary 1965 plan presented by the chief of engineers included very generalized specifications for the lower-river hurricane-protection plan. In sketchy terms, Congress authorized a plan that called for improving the existing but inadequate back levees and making adjustments to the river levees as necessary to bring them into alignment with hurricane design needs.[1] The design standard was to protect against "tides of 100-year frequency" or a storm with an average wind speed of 91 mph. This was a lower standard than used on the more urbanized Lake Pontchartrain and Vicinity components, and the design memorandum indicated the project "will not provide complete protection from tidal flooding."[2]

On the East Bank of the river, hurricane protection would extend downstream to Bohemia and on the West Bank to Venice (see fig. 5.1). Following Hurricane Betsy, the Plaquemines Parish Commission Council requested permission to expedite construction of the levees between Phoenix and Bohemia, the downriver segment. By September 1968, it had completed the first lift of this section to a grade of fourteen feet above mean sea level (msl). These levees successfully offset serious damage when Camille blew past in 1969. Nonetheless, that storm produced serious flooding on the West Bank of the river and caused massive destruction south of Port Sulphur.[3]

The lower delta provides only a thin line of viable terrain for human settlement. A string of small communities and farms cling to the river levees on each bank of the waterway. In 1970, Plaquemines Parish had more than twenty-five thousand residents. It had witnessed population losses owing to Hurricanes Betsy and Camille, but the Corps' economic assessment projected rebound by the late 1970s and also economic growth in the parish.[4] Residential, urban, industrial, commercial, and agricultural activities clustered atop the narrow natural levee. Elevation of the natural levee is about five feet above sea level at the upper end of the project area and grades to near sea level where the road ends at Venice. A pair of highways, one on each side of the river, connects the string of communities. Some agriculture, primarily citrus and livestock, takes advantage of the location's long frost-free season. Fishing and fish processing are prominent commercial activities, along with oil-and-gas-related infrastructure and industrial activity dominated by the sulphur-processing facility at Port Sulphur.

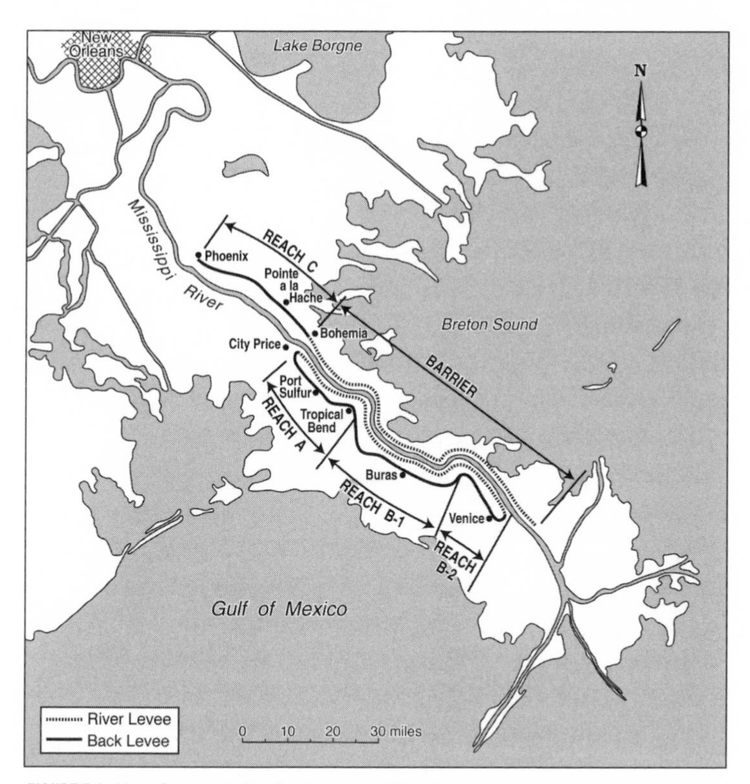

FIGURE 5.1 New Orleans to Venice levees and barriers

As part of the delta, subsidence in the New Orleans to Venice territory was a critical concern for building massive earthen levees. Subsidence rates ranged from 0.5 to 1 feet per century along the upper sections to 5 feet or more per century on the margins of the delta.[5] Much of the land area in this project territory was marshy delta terrain created by the river before the flood-protection levees began diverting rejuvenating sediments into the deep gulf. Deprived by the river levees of new sediment, both the marsh and the natural levees are subject to net subsidence.

Reach A—City Price to Tropical Bend (West Bank)

Located at the upper end of the West Bank construction zone, Reach A achieved completion much later than the other New Orleans to Venice sections. An initial design memorandum for the reach appeared in 1983, but consideration of a high-strength geotextile material to reduce costs and minimize levee weight delayed construction. In order to evaluate the innovative material's effectiveness, the Corps built a test levee using the geotextile material. After several years, the test satisfied the Corps that the material could strengthen the levee without adding unnecessary bulk and weight. Ultimately the decision to incorporate the new technology led to a major revision of the design for Reach A, with an improved factor of safety.[6] Nonetheless, this process set back the completion of a design memorandum and construction over a decade.

The City Price to Tropical Bend reach is a narrow stretch of natural levee in the lower delta. Several small communities and a mixture of maritime, fishing, oil, and agricultural land uses predominate the 4,300 acres in the reach. A locally-built back levee provided limited protection from hurricane flooding in 1987. The Corps' hurricane-protection plan called for constructing 12.8 miles of back levees with an elevation ranging from 11 to 14.5 feet that would tie into the Mississippi River levees. A design storm with maximum winds of 85 mph and a forward speed of 12 mph provided the basis for this design.[7]

In a setting dominated by human manipulation of the landscape, the Corps' environmental assessment concluded that there would be minor annoyances owing to dust and noise as trucks hauled material to build the levee. It also projected some longer-term environmental impacts owing to conversion of wetland habitats to levees or borrow pits. Nonetheless, the impacts resulting from increased hurricane protection, the Corps observed, would yield an overall positive result.[8]

Use of the geotextile material in construction, planners forecast, would result in $48 million in savings. Overall, the Corps calculated that the Reach A project cost $4.5 million and yielded more than $10 million in benefits; thereby producing a positive benefit-to-cost ratio.[9]

Reach B1—Tropical Bend to Fort Jackson (West Bank)

The Tropical Bend reach, on the West Bank, extends from just upstream of Empire to Fort Jackson. It is the central reach of the West Bank component and occupies a narrow band of relatively high ground along the river's natural levee. Its population growth rate stalled dramatically following the two hurricanes of the 1960s, but growth crept ahead nonetheless. Most homes in the parish employed stilt construction to raise the dwellers and their belongings above frequent hurricane flooding. This provided some protection when surge overtopped the levees and offered some safety to those returning after hurricanes Betsy and Camille. Projections of annual flood damage without the hurricane-protection system were gloomy. For the Tropical Bend Reach, the Corps estimated that the back levees would prevent some $3.8 million in damages to both current and future development. Further justification for the project came from the observation that, without protection, the area was entirely unsuited to development and human residence, and since it was already occupied, levees were essential. Again, the circular logic that levees justified development and development justified levees appeared. Overall, the Corps' calculations showed that the authorized back levees would yield nearly $4 million in benefits.[10]

The 1971 design memorandum contained significant alterations from the earlier 1967 version. The latest iteration substituted sand-core levees capped with clay for all-clay levees. This took advantage of locally abundant sand supplies, while minimizing the problem of inadequate local, suitable clays. Also, based on recommendations from the parish governing body following Hurricane Camille, the Corps shifted the levee alignment closer to the river and thereby reduced the costs the Plaqeumines Parish Commission Council would have to bear. Using a one-hundred-year frequency as the project storm, the design called for levees with a net grade of fifteen feet.[11]

The U.S. Fish and Wildlife Service commented on the proposed plan and expressed concern that hunters and fishermen might lose some access to wetlands. It also noted that adverse effects on wildlife "will depend on the manner in which dredging, spoil handling, and spoil disposal is accomplished." The wildlife agency recommended that the Corps take every measure to preserve habitat.[12]

Reach B2—Fort Jackson to Venice (West Bank)

The Fort Jackson to Venice reach is another narrow stretch of natural levee with a linear settlement of small citrus farms and homes of families engaged in fishing and fish processing and oil-related activities. Venice is the largest community on this stretch and is home to a considerable fishing port. The project area encompasses some 2,300 acres. Planners used an average maximum wind speed of 96 miles per hour to establish design criteria.[13]

Levees in the B2 reach would employ the sand-core, clay-cap construction technique used in the B1 reach. Access to large quantities of sand pumped from borrow areas in the Mississippi River provided ample quantities of this core material. Construction crews would build these levees to an elevation of fifteen feet.[14]

The Corps' environmental assessment concluded that mining clay for the levees and the construction of spoil and ponding dikes, along with the removal of upper-layer clays to gain access to desirable building materials, would not impact nearby commercially important oyster beds. Corps officials acknowledged short-term impacts owing to turbidity produced by clay removal and storage but concluded that restoration efforts following completion of the project would balance out any short-term damage. The design memorandum also noted that hurricane protection and the encouragement of orderly development within the levee system would provide substantial benefits outweighing any detrimental impacts.[15]

Considerable cost increases from the initial project design exceeded those of other reaches. Of the more than $20 million increase, $17 million resulted from revamping the net grade of the levees. Nonetheless, the Corps calculated that the project would produce an average annual benefit of slightly more than $1 million. Compared to an average annual cost of $789,400, the project still yielded a narrow beneficial ratio.[16]

Reach C—Phoenix to Bohemia (East Bank)

The lower-river segment stretching between the communities of Phoenix and Bohemia occupies a similarly narrow band of natural levee immediately adjacent to the Mississippi River. As it did with other segments, the Plaquemines Commission Council initiated construction of interim levees to a height of fourteen feet soon after Hurricane Betsy and completed that first stage by 1968. The 1972 Corps plan called for raising the interim levees to seventeen feet. This height was to protect against a storm with an average maximum wind speed of 96 mph and a return frequency of 100 years.[17] As with the other lower-river hurricane-protection levees, the Reach C segment would provide a back levee that would loop around the gulfward side of the natural levee and tie into the preexisting river levees. Gravity drainage would collect rainfall within the bowl-like levee ring, and two pumping stations would augment the gravity-fed drainage network by expelling the runoff. Local interests had provided the drainage system. In addition, they constructed the interim levees using the sand-core, clay-cap technique.[18]

Corps planners noted that "the only potential damage to the adjacent ecosystem would be as a result of the erosion of fill material by rain or material drift due to wind action." They also pointed out that a short- and long-term benefit to the human ecosystem would be protection from hurricane-induced flooding.[19]

Overall, Reach C would protect some 4,500 acres. Corps calculations indicated average annual costs of $461,900 compared to average annual benefits of $870,000, and a favorable benefit-to-cost ratio.[20]

BARRIER FEATURES

In addition to the loop levees included as part of the initial plan, the Corps proposed what they referred to as "barrier features" for both the East and West Banks of the river (see fig. 5.1). On the West Bank, the barrier feature would be an enlarged Mississippi River levee with an elevation ranging from 13 to 15 feet. The East Bank barrier would consist of a 15.8-foot levee along the East Bank of the river from Bohemia to a point just below Venice. The basic purpose was to impede easterly storm surge from moving across the delta or from the river to the West Bank. An environmental impact statement in 1975 provided the initial assessment of the barrier features that met stiff resistance from environmental organizations. In particular, questions surfaced about the impact of the East Bank Barrier Levee. It effectively would sever the river from "one of the last remaining functional alluvial levees in Louisiana." If constructed, the barrier would eliminate overbank flooding, hold sediment rejuvenation of the bird's foot delta, and accelerate subsidence.[21] The environmental questions clouded the desirability of the East Bank barrier.

Construction of New Orleans to Venice Hurricane Protection

The Corps' final environmental impact statement noted that there would be short-term environmental disruptions during construction but concluded that since modest back levees were already in place, it was difficult to identify any viable alternate plans. With construction merely to add to the height of the existing levees, the environmental impact statement reported that environmental impacts would be minimal.[22] The Corps' EIS justified construction with the argument that, in the absence of the existing human settlement, there would be no need for this hurricane protection system but the presence of people validated construction and the minor environmental impacts. It also concluded that the planned hurricane protection system would interrupt "a dynamic system," but it pointed out that if the dynamic system were allowed to proceed, it "would result in the destruction of this area [the human landscape]."[23] It acknowledged that the existing levees were already preventing the creation of a fan-shaped delta by directing sediments offshore. The Corps also observed that "the construction of the levee system may promote a false sense of protection which could result in the relaxation of building and construction codes by local government."[24] To counter this tendency, the Corps recommended local governments enact stringent building codes as a prerequisite to construction. Overall, the Corps indicated the hurricane protection for the entire area offered sufficient benefits to offset the other environmental issues.

Consideration of both the continued diversion of sediments into the gulf and the false sense of security illustrates that the Corps was aware of key issues of survival in the delta region. Yet, the river levees had been in place for many years, and the hurricane-protection levees would reduce the tropical storm damage to an existing population. The purpose of the EIS was to assess environmental impacts of a particular hurricane-protection design, not to recommend major social and economic adjustments. There were no objections to the EIS along the lower delta, and approval cleared the way for construction.

Following environmental clearance, and eventually completed designs, federal work on the lower delta hurricane-protection system proceeded after some delays over right-of-way. It took just over two years to conclude agreements over rights-of-way for Reach B1. In the meantime, local interests completed Reach C by 1968, and federal work on Reaches B1 and B2 began in the early 1970s.[25] By 1976, first lifts on Reaches B1 and B2 were nearing completion.[26] Crews had finished the first lifts for B1, B2, and C by 1982.[27] As of 1985, the Corps reported that work was 32 percent complete on the lower delta projects.[28]

Local interests at the time requested that the Corps defer work on Reach A and direct all effort toward the East Bank Barrier, which had not received approval of its environmental-impact statement. Following the EIS process, the Corps and local interests dropped the East Bank Barrier and instead advanced an alternative plan that would provide just a West Bank Barrier and a mitigation plan to provide a natural marsh on the East Bank. Both parties finalized agreements on this option in late 1987. Additionally, in 1987, after the geotextile fabric tests provided favorable results, the Plaquemines Parish government requested that the Corps resume work on Reach A.[29] Some 42 percent of the project was completed by 1989, and construction approached the two-thirds mark by 1992.[30]

In order to allow local residents to take advantage of the federal flood insurance program, the parish government requested the Corps upgrade an additional 3.3 miles of levee at the upper end of Reach A. The parish offered to pay for this construction, and the Corps approved the project in 1992. By 1997, the Corps' crews had nearly completed three quarters of the overall project.[31] Project Manager Carol Burdine reported in 1999 that the New Orleans to Venice project was 79 percent complete.

Contractors faced unique challenges working in marshy conditions. Nonetheless, construction crews in 1999 completed a second lift to several miles of levee near Port Sulphur.[32] Following Hurricane Lili in 2002, the Corps ordered parish crews to halt work on a levee to protect the parish's only highway. The parish had begun this work without an environmental permit, and it claimed that the Corps was putting wildlife ahead of public safety as the 2003 hurricane season approached.[33] Progress slowed in the new century, with construction reaching only the 80 percent completion mark in 2004.[34]

Congressional appropriations for the New Orleans to Venice component did not match the sustained support seen for the Lake Pontchartrain and Vicinity. After 1984, the initial work allowances provided by Congress were below the amount requested in the president's budget. Although the shortcomings were not substantial enough to derail construction, they were consistently 80 to 90 percent below requested funding.[35] Delays crept into the process as the Corps examined the geotextile liner and dealt with the environmental impacts of the East Bank Barrier. Complaints about funding diversions for the entire lower Mississippi River levee-building projects emerged in 2004. According to Brigadier General Don Riley, commander of the Lower Mississippi Valley Division, 2004 marked the third year in a row that the division's budget declined. Senator Mary Landrieu released figures showing consistent budget reductions from 2002 through 2005 for hurricane protection across Louisiana. In 2003, Congress appropriated less than half the money requested by the Corps for hurricane-protection projects. In 2004, Congress appropriated 27 percent of the Corps's request and in 2005 only 21 percent.[36] Late in the 2004 fiscal year, the New Orleans District reported, "For fiscal year 05, the President's budget is $2.97 million, and our capability is $6.6 million. Funding constraints will continue to slow down work on the project and extend the project completion date."[37]

LONG DELAYS ON THE WEST BANK

Within the New Orleans urban sprawl is a complex suburb known as the West Bank in Jefferson Parish. It is situated primarily on the traditionally habitable natural levee on the West Bank of the Mississippi River. Separated from the East Bank portion of the parish by the river and its massive levees, socially and economically the West Bank is very much a part of the metropolitan area, but it functions as a separate drainage basin and requires its own drainage and hurricane-protection systems. The natural levee extends a mile or two back from the river and follows a gentle gradient from 12 feet above sea level to the marsh, which grades from 1 to 0.5 feet above sea level. Land within some of the locally-built levee-enclosed tracts had subsided several feet below sea level in the early 1980s. The Corps projected continued subsidence of 1 to 2 feet in leveed areas, coupled with sea level rise of 0.5 feet over the next century.[38]

Industrial land uses occupied considerable portions of the West Bank riverfront. Shipbuilding, grain storage, and petrochemical processing are some of the prominent activities. Numerous small towns have coalesced into suburban sprawl; Gretna, Harvey, Marrero, and Westwego are the most prominent communities opposite New Orleans. In order to accommodate an expanding population, Jefferson Parish had extended its drainage districts off the natural levee to create new suburban tracts in former wetland areas on the West Bank before 1970.[39] Jefferson Parish's population was only 53,441 in 1950 but had grown to more than 200,000 by 1960 and continued upward to more than 450,000 by 1980. With an enlarged territory claimed from the surrounding wetlands, the West Bank's population soared from 32,000 in 1940 to 180,000 in 1980. Although distant from the Gulf of Mexico, those residing on the West Bank were vulnerable to hurricane flooding. Storm surge can move inland across the wetlands south of New Orleans and penetrate the metropolitan area via canals and other waterbodies. The West Bank's low-lying position exposed unprotected zones to inundation. When Hurricane Betsy swept over New Orleans and vicinity, the West Bank was in a critical growth phase and susceptible to flooding. Although the West Bank escaped the serious flooding in 1965, residents and officials there sought to share in the improved hurricane protection discussed after Hurricane Betsy.[40]

The Corps of Engineers considered including the lower areas of West Bank Jefferson Parish in the New Orleans and vicinity hurricane-protection system. Representatives from the Corps met with local organizations and individuals in December 1966 to hear their views. Following that meeting, Corps staff conducted evaluations that indicated economic factors did not warrant hurricane protection for the rural bayou communities of Barataria and Lafitte—in southern Jefferson Parish. Planning for the suburban portions of the West Bank continued, however. In July 1972, the Corps presented several optional levee alignments for the West Bank. Local developers were unsatisfied with the small footprint depicted for protected areas and argued for extending the levee system farther from the river into the wetlands, thereby offering the parish more expansive development opportunities and an enlarged tax base. Local environmental groups opposed the prodevelopment position and supported a levee alignment that would encircle only the existing urbanized area. Further complications arose over the relationship of the levee system and a national park that was in the early planning stages. Seeking to minimize the construction of levees within the park, Corps officials, in 1973, delayed investigations until the National Park Service finalized the boundaries for the Barataria Unit of the Jean Lafitte National Historic Park and Preserve (see fig. 5.2).[41] Final demarcation of the park's territory did not occur until 1979.

This was just the beginning of a protracted process that greatly delayed progress on the West Bank hurricane-protection system. Following its 1973 decision to delay West Bank planning, the Corps encountered additional situations that impeded progress. According to the Corps' own feasibility study, "inadequate funding, work load increases, shortage of personnel, and implementation of new guidelines which imposed additional economic, engineering, and environmental study requirements" on its staff held up progress.[42] In 1977, the Corps resumed work on planning and, by May 1978, presented fifteen alternate levee alignments to the Jefferson Parish Council. After considering the Corps' proposal, the council countered with its on alignment that would have extended coverage around Crown Point, an area encircled by an isolated levee ring that was well beyond the contiguous existing protection barrier. An economic evaluation suggested that this option did not achieve a positive benefit-to-cost ratio.

Finally, after congressional approval in 1978, the National Park Service (NPS) authorized the creation of the Barataria unit of the Jean Lafitte National Park and established its boundaries in 1979. This enabled parish officials and the park service to select a preferred levee alignment that avoided intruding on the park but also maximized West Bank protection.[43] As an expression of local anxiety, the Jefferson Parish Council passed a resolution that same month asking if the Corps could expedite the hurricane-protection project.[44]

With the configuration that avoided the park in mind, the Corps proposed a federal alignment to the Jefferson Parish Council in September 1979. Apparently dissatisfied with the limited territory encompassed by the Corps' plan and its slow progress, the Jefferson Parish Council passed a resolution to decline federal funds and to build the hurricane-protection levee system on its own. Council members believed the parish could build the levee faster and cheaper than the Corps—and with greater tax benefits to the parish.[45]

The parish's desire to accelerate construction proved ill conceived and encountered criticism from multiple fronts. Local environmental groups chastised the parish for delaying progress by wresting the project from the

Corps and thereby endangering its constituents.[46] Westwego mayor Ernest Tassin opposed the Corps' alignment, claiming it would displace citizens in his community. Local developers claimed the Corps' proposed levee imposed a "no growth" line that would stymie economic development in the parish.[47]

The parish responded to its critics and claimed that by taking over the project and rejecting federal funds, it could avoid delays associated with environmental impact statements and other requirements the project faced when relying on federal dollars. Yet, when the parish council submitted a permit application to proceed with its levee alignment in June 1981, the Corps, which considered applications of this sort, ruled that the parish, contrary to its expectations, would have to conduct a proper environmental impact statement. Unfamiliar with the process, the parish took several years to complete its EIS. By April 1984, Jefferson Parish presented its EIS to the Corps, which convened a public meeting to hear comments on the proposed plan and its environmental consequences. There was a spirited discussion at this meeting, but ultimately the public debate had little bearing on the outcome of the parish's plan.[48] After considering the EIS, the district engineer denied the parish's application in June but agreed to permit a more restrictive alignment that did not include the controversial Crown Point extension.[49] The Corps' justification was that the smaller alignment would preserve some 1,940 acres of wetland and would impose less financial burden on the public. Amid reminders that the parish was still vulnerable to a major storm, the parish reconsidered its options.[50] Frustrated with the Corps' decision, the parish council voted to return the project—design, primary financial obligations, and construction—to the Corps in September 1985. Consequently, the Corps resumed planning.[51] Some twenty years after Hurricane Betsy, there had been virtually no progress on the West Bank for a variety of reasons involving the parish, other federal agencies (NPS), and basic procedures.

Hurricane Juan in 1985 highlighted the consequences of these delays. Juan passed slowly along the Louisiana coast south of New Orleans in late October (see fig. 4.3). Although it barely achieved hurricane wind velocities, Juan drifted on a track from Morgan City to Burwood (in the lower delta) pushing considerable surge into the bayous, lakes, and canals south of New Orleans and the West Bank. Persistent high winds produced surges 5 to 8 feet high well inland.

Storm surge reached a maximum height of 4.74 feet in the Harvey Canal, exceeding the previous record (4.2 feet during Hurricane Carla in 1961). While local organizations had been working on hurricane levees along the canal, the local levee district had not completed a critical section and, at the time of the storm, had suspended work on that particular project. High water pushed into the canal overtopped this incomplete portion of the protection system. A prompt response by emergency crews that sandbagged the opening prevented serious residential flooding.[52]

High water also overwhelmed the existing West Bank levees in Westwego. There, surge breached the levee and flooded vacant land. An emergency response by the Corps erected an earthen dike that diverted the floodwaters toward a pumping station, which lifted the water back into the adjacent wetland. Another levee built by a local developer proved insufficient to hold back surge, and overtopping flooded hundreds of homes in the Westminster and Lincolnshire neighborhoods. Water 3 to 4 feet deep forced evacuation of these neighborhoods. In addition, the V-line levee, built by the local drainage authority, suffered overtopping, and eventually a portion collapsed. An emergency response enabled the construction of a temporary ring levee to contain the flood waters. This prevented serious flooding of residential areas.[53]

Overall, Juan produced more of a scare than damage, although residential and commercial flooding captured the attention of local authorities and residents and reinvigorated concerns about hurricane protection. It clearly demonstrated that even a fairly weak storm could overwhelm the existing levee system. This convinced local interests that it was time to accelerate hurricane-protection planning and construction. After years of conflict and delay, parish and state officials demanded that the Corps reactivate its West Bank hurricane-protection planning process.[54]

Westwego to Harvey Canal

With renewed public support, planning for the major sector of the West Bank project moved forward after Hurricane Juan. The 1986 Feasibility Report considered several options and their social and environmental impacts. It recommended the V-south plan based on its favorable economic features and the level of protection it would offer (see fig. 5.2). Although this

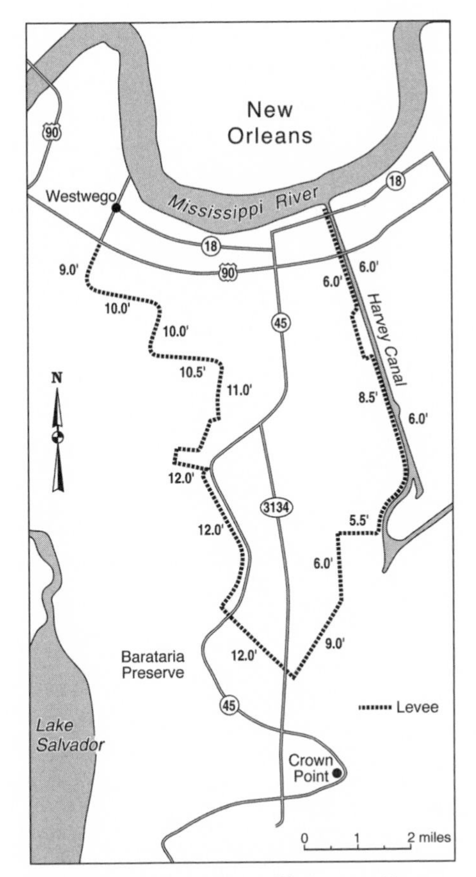

FIGURE 5.2 West Bank levees, Westwego to Harvey Canal. After U.S. Army Corps of Engineers, 2003.

alignment would permit accelerated urbanization of wetlands, the Corps commented that development pressures would ultimately encroach on this area even without a federal project.[55] Once again, officials used what they considered inevitable urban growth to justify their levee alignment. The recommended plan would require congressional approval to encroach on the Bayou Aux Carpes site, an area designated by the U.S. Environmental Protection Agency as a protected wetland. It would also require use of thirty-three acres of land in the Barataria unit of the Jean Lafitte National Park.[56] Economic considerations drove this option, which required additional delays since it hinged on legislative approval.

In July 1989, the Corps released the design memorandum for the Westwego to Harvey component and its V-levee north option. The V-levee north option differed from the 1986 recommended plan by avoiding the Jean Lafite National Historic Park and the Bayou Aux Carpes wetland area and thus avoiding procedural complications. Designers based their work on a standard project hurricane of 100 mph with an estimated return frequency of 500 years.[57] The fundamental purpose of this component was to provide improved levees for the urbanized area of the West Bank between the Harvey Canal on the east and lakes Cataouatche and Salvador to the west.

Based on the standard project hurricane, the Corps calculated levee heights ranging from 9 to 12 feet.[58] Effectively, the West Bank project would encircle the land from the Bayou Segnette Pumping Station to the Harvey Pumping Station. Portions of this project already had locally built levees that did not meet the required height, while other areas required new levees built from scratch. Most of the levees would be earthen barriers with broad footprints, but the plan called for narrow floodwalls in congested areas and around existing pumping stations. Given the tendency of soils in the project area to subside, designs included a settlement factor of 0.5 feet for the floodwalls.[59]

The urbanized nature of the area to be protected produced a favorable annual benefit level of more than $35 million. Average annual costs, even using the more expensive V-levee north plan, were only $13.8 million. Using the Corps' system, this was a favorable benefit-to-cost ratio and justified moving forward.[60]

West Bank—East of Harvey Canal

A second component of the West Bank hurricane-protection system was the area east of the Harvey Canal. It would provide levee protection around the communities of Algiers, Terrytown, Gretna, and Harvey (see fig. 5.3). The proposed levees would follow an existing tidal protection levee and raise it to a suitable level. It would connect the Mississippi River levee at the Algiers Lock with the levee on the east side of the Harvey Canal. This would effectively enclose the urbanized West Bank communities with levees 9.5 feet high on their vulnerable flank facing the southern wetlands. The plan would add two feet of levee protection to the area east of Harvey Canal.

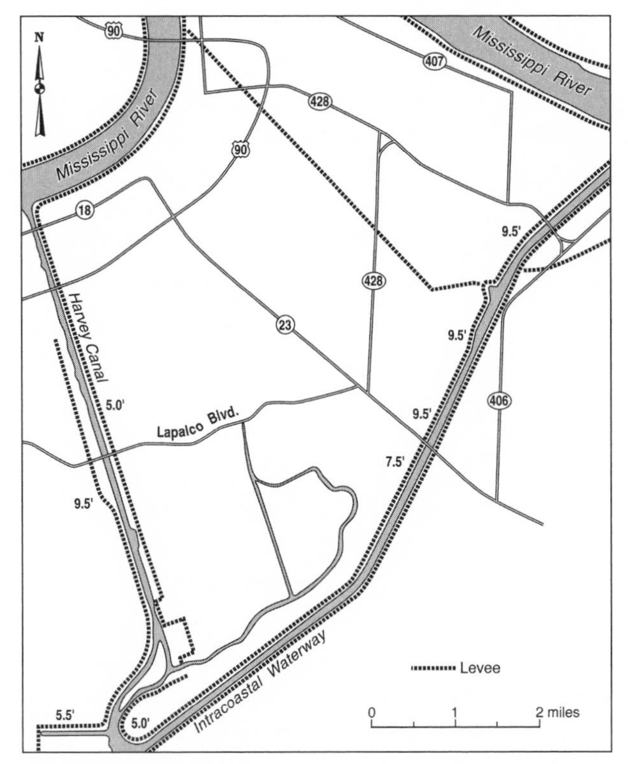

FIGURE 5.3 West Bank levees east of Harvey Canal. After U.S. Army Corps of Engineers, 2004.

Corps planners used a standard project hurricane of 115 mph and a forward motion of 12 mph to calculate the levee's design heights. This was the least-cost plan and had a projected price tag of $25.7 million.[61]

Extending the Protection Upriver

In addition to West Bank Jefferson Parish, the next parish upstream also sought hurricane protection not included in the 1960s planning. Hurricane Juan in 1985 and later persistent southerly winds from Tropical Storm Frances caused flooding in the low-lying sections of St. Charles Parish and prompted the parish to take action. Suburban sprawl in St. Charles Parish had pushed housing developments into the vulnerable backswamps off the natural levee. Local deliberations on an alignment in 1994 failed to achieve a resolution. But in 1998, the parish began building a makeshift levee to hold off surge pushed northward from lakes Catouatche and Salvador into new residential neighborhoods. The parish argued, unsuccessfully, for the Corps to include its new neighborhoods in the hurricane protection plan. Like Jefferson Parish, it envisioned a levee system encompassing a large area of undeveloped wetlands, but the Corps sought a smaller footprint for the protected area—both to limit costs and to minimize encroachment on wetlands. After five years of debate, the St. Charles Parish Council accepted the Corps' levee alignment in 1999 in anticipation of building its own hurricane-protection system.[62]

Progress on West Bank Projects

Long after the substantial delays associated with the local adoption of levee-building responsibility and its subsequent return to the Corps of Engineers, construction moved forward slowly. As of September 1990, the Corps had expressed concern over delays in securing Local Cooperative Agreements with local interests. The initial delays were owing to local sponsor's concern over the cost-sharing arrangements of the cooperative agreement. In addition, local sponsors also sought to protect additional areas east of the Harvey Canal not included in the Corps' levee alignment and to add a levee at Lake Cataouatche. For the Westwego to Harvey Canal section, local interests objected to the unexpected price of cleaning up hazardous waste found in the construction area. The Local Cooperative Agreement mandated they remediate the wastes, but haggling over this item delayed progress. East of the Harvey Canal, industrial interests objected to the levee and floodwall being built along the east edge of the Harvey Canal that they thought would restrict commercial activity.[63] The Corps responded with a revised flood gate and a series of floodwalls and levees that would not interfere with activity along the canal.[64] Additional delays accompanied discussions about cost sharing for the pumps associated with the hurricane-protection system. In spite of the delays, local interests appealed to the Corps to expedite construction. In a June 1991 letter to Colonel Michael Diffley, district engineer, Jefferson Parish president Michael Yenni wrote, "Due to the magnitude of recent flood losses to our citizens, these projects must move ahead as expeditiously as possible. I understand that the Corps' authorization process is very deliberate and time consuming. However, any further delay in the design and construction of these important projects will bring unnecessary suffering and damages to the citizens of Jefferson Parish."[65] Local officials continued to criticize the Corps' slow progress at a public meeting held in July 1994. Ron Besson, former president of the West Jefferson Levee District, complained that the work was moving at a "snail's pace" and that numerous studies were a waste of taxpayers' money.[66]

Further investigations of the hazardous-waste cleanup and the various adjustments to the plan continued during the early 1990s. The project alterations, and the associated investigations required, increased the projected costs, and ultimately the Jefferson Parish partners determined that they would not be able to cover their 35 percent share ($144 million). Louisiana governor Edwin Edwards stepped in and offered to designate the State Department of Transportation and Development as the local sponsor. In a letter to the Corps, he stated that he was "concerned that significant delays will occur in providing hurricane protection to this area, with potentially catastrophic impacts." This enabled the Corps to move forward with planning for the Lake Cataouatche and the East of Harvey Canal projects.[67]

Work was under way on less problematic components by this time, and, indeed, work commenced on several elements of the West Bank project in 1991.[68] Corps officials determined that the project was 10 percent complete by September 1992.[69] Local interests were not satisfied with the pace and urged "the rapid completion of all pre-authorization activities. Until this project is completed, the heavily populated areas east and west of Harvey Canal remain critically exposed to the threat of hurricane flooding."[70] Some of the delays up to that point were owing to complications encountered in completing the draft feasibility study.[71] By 1995, crews had completed 20 percent of the work on the Westwego to Harvey Canal project.[72] In 1998, parish levees failed near the community of Lafitte during the passage of a relatively mild tropical system. These levees were not part of the Corps' project but highlighted to parish officials the fragile situation of West Bank protection. Indeed, this mild storm came within six inches of topping some of the levees east of the Harvey Canal. Delays in this section stemmed from lack of state funding—despite the governor's commitment—to match the appropriated federal dollars.[73] By 2000, the Corps reported that the entire West Bank project was 22 percent complete and estimated the task would be completed by

2014.[74] Despite uncertainties about funding, the Corps awarded a contract on the Harvey Canal floodgate in October 2002.[75] Construction made considerable progress by 2004 when the Corps announced that it had reached the 36 percent completion mark. By May 2005, the project achieved 38 percent completion. In early 2005, crews were just getting started on the east of Harvey Canal and the Lake Cataouatche projects. Corps officials in May 2005 had projected a 2016 completion date.[76]

Even after the Corps and St. Charles Parish agreed on a levee alignment, there were considerable complications. Landowners resisted selling property to the parish and delayed the parish from providing the right-of-way to the Corps. The parish also encountered delays obtaining environmental permits to proceed with the project. In 2002, St. Charles finally received its permit to begin construction of the hurricane-protection project.[77]

Funding for the West Bank project was regular after 1988 when Congress first recommended $1 million for planning. From 1991, the work allowance remained above $3 million annually, peaked at over $20 million in 2000, and saw substantial increases in 2004 and 2005 to $24.3 and $26.6 million respectively.[78] Local funding, however, proved problematic. With declining oil revenues that began in the mid-1980s, both Jefferson Parish and the State of Louisiana faced difficulties meeting their obligations. Parish officials explained their economic plight to the Corps in 1992 and began to push for greater state support at that time.[79] In 1996, local representatives expressed concern that Congress would not sufficiently fund the West Bank project and pressed their congressional delegation to fight for the project's $157 million budget.[80] After Congress made good on appropriations in 1996, the state, which had assumed much of the financial burden for the West Bank projects, took more than three years to allocate its matching funds.[81] By 2000, the parish had made levee construction its top priority and finally received $49.5 million from the state to press ahead with construction.[82] State funding remained questionable during the new century. Despite threats to the levee system during the passage of Tropical Storm Isidore in 2002, Louisiana was unable to fully commit to its 35 percent share. The Corps moved forward with construction nonetheless.[83] Extensive damage from the dual storms of Isidore and Lili in 2002 strengthened the resolve of West Bank officials to fund the hurricane-protection projects. Yet in 2003, local officials charged that there were nonfederal funds available,

TABLE 5.1 West Bank Actual Costs

Fiscal Year	President's Request	Congressional Committee Recommendation	Work Allowance (Enacted)
2004	35,000	28,500	24,326
2003	5,000	9,000	7,338
2002	12,000	12,500	10,502
2001	8,065	8,065	6,759
2000	7,000	15,070	12,924
1999	3,936	7,000	6,509
1998	6,685	7,426	6,794
1997	4,256	7,206	6,818
1996	2,100	2,100	1,732
1995	6,600	6,600	5,551
1994	6,270	6,270	5,366
1993	9,000	9,000	7,947
1992	3,800	3,800	3,238
1991	3,700	4,300	3,792
1990	0	1,100	1,085
1989	600	600	573

Source: U.S. Army Corps of Engineers, New Orleans District, "New Orleans Projects—Actual Costs," unpublished spreadsheet, 2005.

while federal dollars were lacking at the tail end of the federal fiscal year. Despite funding challenges, local officials foresaw no delays to existing contracts.[84] Louisiana senator Mary Landrieu's office charged that congressional funding for hurricane-protection work in Louisiana had consistently fallen far short of the Corps' requests since fiscal year 2002. From 2002 to 2005, appropriations never reach half of what the Corps requested.[85] In May 2005, the Corps reported that the fiscal year 2005 presidential budget of $37 million was more than $20 million short of their capability, and Congress appropriated only $30 million.

PROTECTING GRAND ISLE

Grand Isle is a barrier island situated in the Gulf of Mexico at the southernmost tip of Jefferson Parish. It was about 7.5 miles long in the early 1970s and about three-quarters of a mile wide. High ground stood about five feet above sea level, and a live oak forest covered the island's summit. Dunes near the shore stood up to eight feet high. The island's position in the Gulf of Mexico and its low profile make it subject to frequent and sometimes intense impacts from hurricanes. Combined storm surge and waves have washed over the barrier island on numerous occasions. A powerful storm in 1893 destroyed the grand inn there, which was never rebuilt. Despite its highly exposed situation and devastating impacts in the past, the island, with its expansive beaches, has remained the site of recreational activity and also the home to a resident population of more than 2,200 in 1970.[86]

Hurricane Betsy produced extensive damage when it roared over the island in 1965. Observers recorded wind gusts up to 160 mph when Betsy moved inland. Surge and waves washed over the entire island, destroying most buildings, even those built on stilts.[87] Storm-driven waves eroded the beach dunes and drove oceanfront sand back across the island.

Erosion of the island's gulf shore by ocean currents has been a longstanding issue on Grand Isle. This erosion has encroached on the desirable beachfront and also threatened homes and businesses. Over the years, local interests made several efforts to control the coastal erosion. In 1951–52, the Louisiana Department of Transportation constructed several groins, or barriers perpendicular to the shore, in the gulf to trap sand transported by longshore currents. A subsequent study showed the groins were ineffective, and the state renourished the beach with sand collected elsewhere. Continuing erosion removed much of the supplemental sand, and Hurricane Flossy in 1956 removed most of what remained from the island.[88] Additional private and state groins constructed near the east end of the island trapped sand, but accelerated erosion east of the structures. Natural processes removed thirty acres by 1963. Hurricane Betsy destroyed the groins at the island's eastern end in 1965. Hurricanes Carmen (1974) and Babe (1977) damaged the restored dune as planning for improved protection moved forward.[89]

The Corps recognized that it would not be able to protect Grand Isle fully from hurricane impacts. Rather than rely on levees, it developed a plan that would attempt to absorb much of the energy of storm-driven waves. At the center of the plan was an 11.5-foot-tall dune near the beach. When storms struck the island, this barrier would take the initial wave impact. Although the design assumed a storm would erode the dune, waves would expend their energy removing the human-made dune rather than the fragile barrier island beneath it. The Corps' enhanced dune would minimize wave damage to the area behind it as long as the barrier remained in place. In addition, the plan called for a jetty to stabilize erosion at the western end of the island. For additional residential protection, building codes on the island mandated that structures had to be built on stilts eight feet above ground level.[90]

The Corps' environmental impact statement for the dune-jetty construction project concluded that there would be minor disruptions to beach and littoral biological communities. Nonetheless, "natural replenishment" would occur within a few years, thus offsetting any short-term impacts. The Council on Environmental Quality received the final environmental impact statement in 1976.[91]

In the interim, serious erosion of the western end of the island continued. Without authorization, the Corps could not undertake work to stabilize that critical zone. Local interests, namely the state of Louisiana, authorized funds and constructed a jetty at the west end and placed sand in the area protected by the jetty.[92]

In 1979, the Corps completed its Phase I design memorandum for hurricane protection and erosion control on Grand Isle. It used the standard project hurricane with wind speed of 100 mph and a recurrence frequency of once in 200 years. Such a storm would generate a storm surge of 9.9 feet.[93]

After considering several options and accepting comments from local interests, the Corps selected the plan that combined beach-erosion control and hurricane protection. It included a vegetated sandfill dune with a crown 11.5 feet above sea level and a berm sloping toward the gulf. This barrier would provide protection for a storm with a return frequency of 50 years and 90 percent protection from hurricane-driven gulf waves. The plan also called for a jetty at the western end of the island to protect it from erosion. This plan would have annual federal costs of $482,000 and benefits exceeding $1.8 million. In particular, the benefits would be erosion protection, land-use development, and recreational activity. This produced a favorable benefit-to-cost ratio.[94] Local interests were enthusiastic and immediately began discussions about building a convention center to lure winter visitors to the tourist island.[95]

The key local concern was that the high dune would obstruct the view of the coast, one of the principal reasons for spending time at Grand Isle. Given the frequency of hurricane damage and with several reminders of Grand Isle's vulnerability after Betsy, local populations accepted the barrier dune. The Corps plan accommodated these concerns by including walkways to traverse the dunes and to ensure access to the beach and protection for the vegetation planted to stabilize the dune. Local interests also expressed a desire for more permanent protection, such as an offshore breakwater, but ultimately they accepted just the dune and jetty plan. Additionally, Louisiana sought and received credit toward its cost-share for the emergency construction of the jetty at the west end of the island.[96]

Once local assurances and construction contracts were in place, offshore sand dredging moved forward expeditiously. Local interests provided rights-of-way and their cash contributions in 1983. Dredging operators signed contracts in 1985, and by September of that year, they had completed the dredging and dune construction.[97] Construction of the entire project was completed in 1991. In advance of Hurricane Lili in 2002, the Corps quickly built an additional 2,500 feet of fabric-wrapped levee, and local residents praised this emergency construction after the surge subsided. Although surge rose as high as four feet in many homes, damage was not as extensive as it might have been without the additional protection.[98] In March 2005, the Corps completed an effort to renourish the gulf-facing dune and beach. In addition, a study was under way to consider protection works for the island's north shore facing the coastal estuary.[99]

LEVEES ALONG BAYOU LAFOURCHE

Communities along the lower course of Bayou Lafourche fell into the Larose to Golden Meadow hurricane-protection project. Bayou Lafourche is a former distributary of the Mississippi River that drains into the Gulf of Mexico just west of Grand Isle. Although severed from the Mississippi River

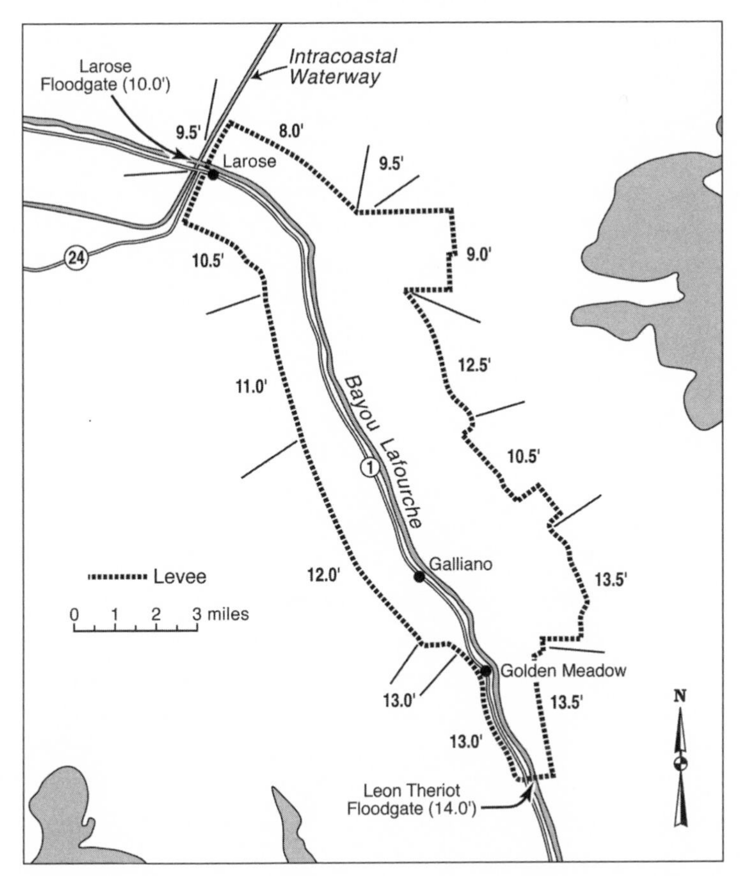

FIGURE 5.4 Larose to Golden Meadow hurricane protection. After U.S. Army Corps of Engineers, 2003.

decades ago by river levees, it carries a small quantity of river water introduced to its headwaters near Donaldsonville via artificial means. Approximately 17,000 people clustered along a rather narrow natural levee in the area where the Corps planned to build levees in 1970. Gulfward of Leeville, the bayou flows through marsh with insufficient natural levee to support urban or agricultural settlement. Sugar cane farming dominates the solid ground along the bayou's upper course above Larose. Below Larose, linear settlements, with a few urban nodes and little agriculture, predominate. Fishing and oil extraction dominate the economies of the communities along the lower bayou. The region is susceptible to storm surge owing to its low-lying situation and the access for surge provided by Bayou Lafourche and the coastal marshes.

The Corps began planning the Larose to Golden Meadow component of the hurricane-protection system soon after authorization in 1965. The initial design memorandum appeared in May 1972 with an anticipated October 1972 completion forecast for the environmental impact statement. Engineers selected design parameters to accommodate a storm with "a 100-year stage" with a wind speed of 89 mph and forward speed of 12 mph. With conditions of this magnitude, the Corps calculated that storm surge would reach 10.2 feet at Golden Meadow with waves of 2.8 feet. In accord with those calculated water levels, the Corps determined that levee heights should be 13 feet high at the southern end. With increased distance from the coast, surge and waves would not be as high, and designers concluded that 8.5-feet levees at the upper end would provide adequate protection.[100]

The plan called for a "loop levee" to encircle the inhabited natural levee territory between Larose and Golden Meadow (see fig. 5.4). This protective barrier was to include both earthen levees and floodwalls where necessary. To prevent surge from moving up the bayou, the plan also called for floodgates on Bayou Lafource at both the upper and lower end of the project area. Planners included floodgates in the barriers to accommodate highways traversing the area. A gravity drainage system would direct accumulations of water to an existing pumping station that would transfer water out of the levee ring.[101]

This plan sought to prevent damage to crops, commercial activities, and residences, and it would encourage future development within its boundaries. Corps planners calculated annual benefits of $4.1 million. Annual costs of construction came in just under a million a year. This produced a highly favorable benefit-to-cost ratio, augmented by strong local support for the project which minimized delays.[102]

Local interests requested modifications to the original plan to include additional private property. With this expansion approved, Corps planners anticipated more wetland losses and made additional modifications to mitigate those impacts. The mitigation plan involved levee construction around a wetland area to maintain freshwater levels and to obstruct saltwater intrusion into the protected wetland. This added costs and time to the project.[103]

Construction of the project commenced in 1975 and reached 31 percent completion in 1982.[104] By 1985, crews had achieved 49 percent completion and 59 percent by 1989.[105] Status of the project reached the 83 percent complete mark by 1994.[106] Progress slowed as levees had to settle, but, by 2000, crews were making second and third lifts on the levees and reached the 90 percent mark. Although the bulk of the key hurricane defenses were in place, additions to the project provided new tasks, and work continued on these elements in 2004.[107] As of May 2005, engineers were preparing plans for third lifts to several levee segments and completing designs for a lock, and they reported the project was 96 percent completed. Corps planners projected a completion date of 2007.[108]

DEVELOPMENTS TO THE EVE OF KATRINA, 1990–2005

The hurricane-protection project lumbered along during the 1990s and the early 2000s. There were a few storms during that period to prompt pleas for acceleration of the deliberate pace. Yet in the decade and a half before Katrina (2005), public officials, the press, and academic researchers issued clear warnings about the continued hurricane risk in the New Orleans area. Disastrous forecasts aided officials who urged massive evacuations in early years of the new century when storm frequency picked up again. As the hurricane-protection project continued to creep forward, an entirely new initiative took form. In the mid-1990s, in response to a spate of heavy downpours, Congress assigned the Corps a massive urban drainage overhaul. Given the fact that hurricanes could cause intense downpours, this project, while separate from the hurricane works, complemented it but also consumed time, talent, and money. Design and construction of the drainage works were largely complete in 2005—an indication that projects could reach completion.

Nonetheless, components of the hurricane-protection works remained unfinished. Work progressed slowly between 1990 and 2005. Several factors account for this measured pace. Final levee lifts required pauses between active construction. Other components required adjustments, refinements, or entirely new components, all delaying completion. Final decisions about the outfall canals enabled construction of parallel levees to begin in Orleans Parish only after 1990. Budget issues, both from the federal level and local sponsors, contributed to the slowdowns. In addition to construction, the Corps was active in developing plans for evacuation and hurricane response. These activities underscore the Corps' recognition of the limits of structural hurricane-protection systems and its obligation to help prepare solutions for more extreme events.

ERRATIC STORM SEASON, 1990–2004

As the hurricane-protection system inched forward during the 1990s, southeast Louisiana did not have frequent tropical storms to compel a sense of urgency. Although most of the major delays had already been put behind them, the Corps and local organizations faced lingering effects from those slowdowns. Most notably, the Lake Pontchartrain system was far from complete, as were most other components of the hurricane-protection system. The metropolitan area was not fully secure from cyclonic storms—even within the design limits of the hurricane-protection system. The infrequent storms that tested the system did not overwhelm the levees, but each reminded local and federal officials that the task was not complete.

Hurricane Andrew was a powerful storm that blew across southern Florida from the Atlantic Ocean in late August 1992, regained wind speed as it crept across the Gulf of Mexico, and unleashed powerful winds across southern Louisiana (see fig. 4.3). Before striking Florida, Andrew had grown into a category 5 storm. As it made landfall on August 24, the National Weather Center reported that the top sustained wind speeds were about 145 mph—a category 4 storm. The storm produced surge along the shore of Biscayne Bay of over 15 feet, although the high water impacted a relatively narrow stretch of coastline. The most extensive damage in south Florida resulted from wind damage to homes. New subdivisions in the Homestead neighborhood south of Miami endured some of the most extensive structural failures. Initial estimates placed damages in south Florida at more than $25 billion.[1]

After passing over Florida and entering the Gulf of Mexico, Andrew's wind speeds crept upward again. It traveled westward across the gulf and then turned northward toward Louisiana. On August 26, Andrew moved toward the central Louisiana coast with category-4-intensity winds in excess of 130 mph. Low pressure and winds created a storm surge up to 8 feet high across coastal Louisiana from Lake Borgne to Vermillion Bay. In the New Orleans vicinity, with the eye well to the city's west, lakefront locations measured surges over four feet, and backswamp areas in West Bank Jefferson Parish saw water rise to similar heights.[2] As Andrew moved inland on a northward track, it weakened, but still left substantial wind damage to inland parishes. It also spawned a fatality-producing tornado inland near the community of Laplace—west of the New Orleans metropolitan area—and several other tornados well inland. Louisiana sustained only one indirect fatality, largely owing to an effective evacuation that mobilized about 1.2 million residents to escape low-lying areas. Andrew's massive impact in Florida worked to convince many Louisiana residents to evacuate with all due haste. The storm dropped rainfall totals ranging from 5 inches to more than 11 inches in New Orleans. All totaled, the 1992 damage estimates for Louisiana stood at $1 billion.[3] Damage to the state's marine fishery and to agriculture were pronounced. Offshore oil production also took a severe blow. Some 3,000 residences suffered damage, but the main urban areas protected by levees recorded no serious damage owing to surge.[4]

Andrew's impact, although more severe in Florida, forced Louisiana and federal officials to see it as an urgent reminder to complete unfinished hurricane-protection business. In the New Orleans area, winds toppled large trees, which disrupted power and damaged homes. Morgan City endured some of the worst damage from winds.[5] Following the storm, state officials reported making hasty improvements to emergency communications systems, reparing evacuation centers, repaving roads and replacing bridges, and also installing a new radar system to track incoming storms.[6] These were short-term fixes to nonstructural systems and offered to help minimize loss of life until the levees were in place.

One of the significant breakthroughs following Andrew, in terms of hurricane protection, was the insertion of disaster-relief appropriations to upgrade the Mandeville seawall. Despite years of public resistance to the planned federal hurricane-protection system, a storm surge of 4.6 feet at Mandeville convinced the city to accept relief funding that enabled the Corps finally to raise the seawall two feet.[7] Additional hurricane protection efforts followed damages on Grand Isle after storm surge and waves washed over the barrier island. Severe erosion of the Corps' built barrier dune prompted repair work completed in early 1994.[8] The fact that the storm's eye passed west of the river delta prevented serious flooding in the highly vulnerable New Orleans to Venice sector.

Hurricane Danny in 1997 provided the next storm of concern for the New Orleans vicinity. It barely reached category 1 status (74+ mph) just before it made landfall on July 19. Tracking northeast across the lower delta with winds of about 80 mph, it quickly decelerated to a tropical depression as it moved across the Mississippi shore (see fig. 4.3). Gusts up to 90 mph blew over Grand Isle, Louisiana.[9] Danny produced damage within the New Orleans District, but impacts were greatest outside the city. Winds damaged boats in the Plaquemines Parish harbor at Buras and surge produced flooding at Empire. In addition, northerly winds at Grand Isle had little effect on the south-facing barrier, but they produced damages on the estuary side of the barrier island. The storm's impact on New Orleans proper was minimal.[10]

Hurricane Georges delivered the next serious threat in September 1998. In route from the Caribbean to the gulf coast, it traveled over the islands of Hispanola and Cuba and raked the Florida Keys. Traveling northwest across the Gulf of Mexico, it generated category 2 winds, although winds decelerated as it approached the Louisiana coast. At landfall near Biloxi, Mississippi, it had sustained winds of about 100 mph. These winds drove the storm surge to nearly 9 feet at Point a la Hache, Louisiana, on the lower delta.[11]

The New Orleans area hurricane-protection system fended off serious flooding, while calls for a mandatory evacuation proved unnecessary as the storm took a less threatening eastward course. Winds blew across Lake Pontchartrain from the north and caused serious damage to exposed lakefront structures, especially camps and restaurants. Floodwaters forced the closing of the lakefront drive between the old seawall and the post-Betsy hurricane-protection levees. There were no reports of serious flooding within the perimeter of levees.[12] Surge caused silting of the Mississippi River Gulf Outlet and disrupted commercial traffic.[13] Wind and waves seriously eroded the Chandeleur Islands—a barrier island complex off the southeast coast of Louisiana.[14] Discussions of reviving the barrier plan and the construction of movable barriers to block surge from entering Lake

Pontchartrain took place after Georges, but high cost estimates deterred full-fledged planning.[15]

In 2001, a slow-moving tropical storm draped massive amounts of rain across the Gulf Coast from Houston to New Orleans. In a week, it unleashed over 20 inches of rain within New Orleans's combined river and hurricane-protection levee system. The deluge, although huge, fell over an extended period of time and did not completely overwhelm the drainage system that was still undergoing improvements. It produced minor street flooding but no major disruptions or property damage.[16]

A more rambunctious phase of extreme weather began in 2002. Tropical Storm Isidore moved northward across the gulf from the Yucatan Peninsula in September 2002. On September 26, it made landfall just west of Grand Isle with wind speeds of about 60 mph. The center of the storm passed almost directly over New Orleans. It produced torrential rains in the urban area, with some locations receiving over 20 inches as the storm tracked inland. Wind speeds were insufficient to drive a storm surge that could overwhelm the New Orleans area hurricane levees.[17] Nonetheless, rain-induced flooding produced troubling consequences. In New Orleans proper, runoff flooded a low underpass on Interstate 10, a key corridor for hurricane evacuation, for twenty-four hours.[18] In addition, floodwater rose to within inches of the top of the Intracoastal Canal levee in Algiers and prompted an emergency sandbagging to prevent serious damage.[19] While neither event caused loss of life or property damage, they revealed lingering weaknesses in the city's defenses should a major storm arrive.

Only a week later, Hurricane Lili made landfall along Louisiana's central coast as a category 1 storm (see fig. 4.3). As it moved inland on October 4, it produced considerable wind damage, blowing over trees and ripping roof materials from houses. Winds and pressure produced a storm surge of 8.3 feet at the Rigolets and at Gulfport, Mississippi. Rainfall totals in Louisiana reached 8 inches in some locations. The combined impact of surge and rain produced some street flooding in New Orleans, but houses in the metropolitan area largely escaped inundation. There was flooding in the Jean Lafitte, Crown Point, and communities in southern Jefferson Parish. Most notable, surge overtopped levees in Plaquemines Parish. This produced flooding of the principal highway and hampered the efficient return of evacuees—once more exposing weak links in the structural system.[20]

Despite the threat of flooding at Algiers during Isidore, the state refused to approve bonds to secure funds as its match to the West Bank Hurricane Protection project shortly after Hurricane Lili. State officials claimed that Jefferson Parish needed to rank the levee work as a higher priority if it expected to secure state financial support. Not only was the state facing a tight financial picture, the Corps of Engineers requested $25 million for the levee project, but Congress recommended only $10 million.[21] Funding had become a critical issue at both levels.

The following year (2003), Tropical Storm Bill moved northward across the Gulf of Mexico from the Yucatan Peninsula. On June 30, it traversed the coastal area of Terrebone Parish on a northeast track near the community of Cocodrie, Louisiana. Wind speeds approached 60 mph but decelerated after the eye moved on shore. Winds downed trees, and surge overtopped a levee near Montegut and flooded some homes. Storm surges topped 5 feet at Mandeville but ranged from three to four feet across most of southeast Louisiana. Although this did not threaten most of the hurricane-protection system, waves and surge damaged the barrier dune at Grand Isle. Estimates placed Louisiana damages at $22 million.[22]

In September 2004, Hurricane Ivan pushed across the Gulf of Mexico on a track toward the New Orleans metropolitan area. Temporarily on a course toward the city, officials called for an evacuation, and the response was massive. An estimated 600,000 residents of the urban area set out for locations beyond the storm's predicted reach. This exodus produced perhaps the worst traffic snarl in Louisiana's evacuation history, and many drivers reported spending more than eight hours in transit from New Orleans to Baton Rouge—an easy one-and-a-half-hour drive under normal conditions. As the storm turned eastward to make landfall near Mobile on September 16, it spared the New Orleans area from serious damage. Nonetheless, winds over 60 mph swept over the lower delta and produced some damage in the toe of the bird's foot delta. The winds and waves severely disrupted oil production in the gulf.[23] Later that fall, on October 10, Tropical Storm Matthew blew ashore near Grand Isle with winds of about 40 mph. It produced a storm surge of over 5.8 feet, flooded homes in Terrebone Parish, and caused extensive beachfront erosion at Grand Isle.[24] The 2004 hurricane season was a mild precursor to events in 2005.

CONTROLLING FLOODING WITHIN THE LEVEES

As the hurricane levee system facilitated suburban growth in Jefferson and Orleans parishes, sprawl produced impermeable surfaces. Roofs of houses and shopping centers, parking lots, streets, and driveways all shed rainfall and contributed to the volume of urban runoff. Suburban development was most rapid during the 1960s and 1970s and pushed development across Jefferson Parish and eastward beyond the Industrial Canal in eastern New Orleans. Population in Jefferson Parish more than doubled from a modest 208,769 in 1960 to 448,306 in 1980 (see table 6.1), and with it came low density housing within the eastern portion of the parish. West Bank Jefferson also expanded into the former wetlands away from the river. Orleans Parish lost population between 1960 and 1980 but expanded its urbanized territory. Population density fell in the inner city as dispersed, single-family housing swept across the eastern lakefront districts that had been largely unoccupied prior to 1965. Within the levee system in both parishes, where much of the post-World War II development took place atop subsidence-prone soils, storm runoff drained to the low ground that was below sea level. Massive pumping systems had to lift this drainage up and into Lake Pontchartrain and other outlets. Population growth stalled after the oil bust of the mid-1980s, but the impermeable surfaces produced by suburban sprawl and the extensive development within the hurricane-protection areas remained in place.

The New Orleans metropolitan area had been spared major inundations during much of the period of suburban sprawl. Climatologists report that between the late 1920s and the late 1940s was a cluster of years with very severe storms. During the Great Depression and World War II, there was little urban growth. Between 1948 and 1978, however, the reverse was true and the New Orleans area endured fewer severe storms as the city experienced its most dramatic growth. Beginning in the late 1970s and continuing through the early 1990s, a cluster of severe storms pounded the enlarged metropolitan region.[25]

Excessive precipitation coupled with substantially more runoff from suburban districts strained the drainage systems and contributed to troublesome urban flooding in low areas of Orleans Parish. Comparable conditions—increased storm runoff from more frequent storms, plus the increased number of residents—caused flooding in Jefferson Parish as

TABLE 6.1 Population in Orleans and Jefferson Parishes, 1960–2000

Year	Orleans Parish	Jefferson Parish
1960	627,525	208,769
1970	593,471	338,229
1980	557,515	455,592
1990	496,938	448,306
2000	484,674	455,466

Source: U.S. Census

well. The series of intense weather began on May 3, 1978, with a storm that dumped a total of 9 inches, with 5.5 inches of that total falling in two hours, completely overwhelming the New Orleans Sewage and Water Board's pumping capacity. Water rose as high as 3.5 feet in the Broadmoor neighborhood and damaged some 1,700 houses in an area known as the "bottom of the bowl." Citywide, floods damaged more than 71,000 houses. The same storm also overwhelmed Jefferson Parish's drainage system. Officials reported that floodwaters invaded 30,000 houses there. In rapid succession, major storms dumped excessive amounts of water on February 6, 1979 (5 inches); April 2, 1980 (over 7 inches), April 13, 1980 (over 8 inches); and June 10, 1981 (almost 7 inches). Urban flooding from heavy downpours had become a pressing issue.[26]

Residents of Orleans Parish clamored for improvements to the drainage system, and in 1981 voters approved a plan to upgrade the city's drainage system. Popular support enabled the Sewerage and Water Board to commence a ten-year program to enlarge the pumping capacity to remove five inches of water in five hours—double the capacity at the time. Meanwhile, Jefferson Parish voters rejected a tax hike to pay for improved drainage facilities in 1978 and opted to rely on flood insurance to protect themselves from flood losses. With flood insurance claims exceeding $93 million between 1978 and 1982, the Federal Emergency Management Agency (FEMA) filed suit against Jefferson Parish for providing inadequate flood protection and for not being in compliance with National Flood Insurance guidelines. FEMA won the first round of this legal confrontation, but an appeals court decided that since floods were "acts of God," Jefferson Parish was not wholly liable for damages these storms produced. Ultimately, the parish and FEMA negotiated a settlement that called for Jefferson Parish to pay the federal agency $1 million dollars and to bring the parish into compliance with the flood insurance guidelines. In addition, the parish had to develop a flood-reduction plan and improve its drainage and pumping capacity substantially. By 1985, both the city and its suburban neighbor had taken preliminary steps to improve their respective drainage systems.[27]

While initial efforts were under way to combat urban flooding, more storms plagued the area in the late 1980s and early 1990s. An April 1988 storm let loose over ten inches that flooded areas in Jefferson and Orleans parishes. In November 1989, rainfall totals ranging between eight and twelve inches also damaged homes across the urban area. Another storm produced two-day totals in excess of eleven inches in June 1991.[28] All these were mere preface to the 1995 downpour. A massive storm on May 8, 1995, dumped up to twelve inches of rain across the metropolitan area and produced widespread damage (see fig. 6.1). This storm unleashed over twelve inches in only five hours at the airport in its first outburst. Overnight on May 9–10, another sixteen inches fell on the northshore parish of St. Tammany. Widespread residential and commercial flooding occurred; high water forced the closing of interstate highways and the airport; and in some communities, up to 70 percent of homes suffered water damage.[29] These storms were not hurricanes, but that clearly pointed out the threat posed by major rainfalls. The massive hurricane levee system could not protect against the secondary threat dropped from the heavens.

In response to pleas from local leaders, Congress in 1996 authorized the Corps of Engineers to assist with rain-induced flooding protection in Jefferson, Orleans, and St. Tammany parishes. Congress undertook this highly unusual commitment of federal dollars to a local drainage system based on the arguments that this area had suffered more than $1 billion in damages since 1978, and the two major urban areas had made considerable improvements to drainage but were unable to keep pace with escalating damages caused by flooding.[30] In addition, residents had shown an exceptional response to the 1968 National Flood Insurance Program (NFIP). The two-parish area had one of the highest subscriber rates for federal insurance. With all-too-frequent claims filed on these policies, the federal government

FIGURE 6.1 Flooding after the May 8, 1995, downpour. Photograph courtesy U.S. Army Corps of Engineers.

hoped to reduce its expenditures on claims with improved drainage. Between the two parishes, more than 126,000 households had NFIP policies, and more than 54,000 policyholders had filed claims between 1978 and 1991.[31] While unspoken, the fact that a major hurricane could unleash more rainfall than frontal systems or summer thunderstorms made the need for improved drainage all the more pressing.

The Corps' plan, known as the Southeast Louisiana Urban Drainage Project or SELA, called for enlarging the capacity of both the drainage canals and the pumps that lift runoff into surrounding waters (see fig. 6.2). In East Bank Jefferson Parish, the Corps included three canals and one pumping station in their enlargement effort. West Bank Jefferson was to receive enlargement of several canals and the installation of a new pumping station. Plans called for a massive reworking of buried culvert systems and enlargement of most pumping stations in Orleans Parish. Initial budgets allocated $27 million from federal sources, to be coupled with local matching funds. The initial budget estimates appear insignificant when compared to actual cost projects of more than $420 million.[32]

Funding for the SELA project began to move efforts forward in 1997 with design work.[33] But by 1999, disruptions to planning and construction

FIGURE 6.2 Southeast Louisiana Flood-Control Project. Features included (a) enlarged drainage canals and (b) increased capacity pumps. Photos by author.

appeared. Work in St. Tammany Parish slowed when, once again, voters declined to approve a sales tax that would have generated the parish's matching funds. In addition, the local drainage district withdrew as the local partner, prompting Slidell to try to salvage the project. The city claimed that it had insufficient personnel to assist the Corps with obtaining permission to complete its surveys necessary to move forward with final design work.[34]

Meanwhile, work progressed more efficiently in Jefferson and Orleans parishes—at least for a while. By 2000, the Corps reported that the SELA project was 40 percent complete, with most of the construction in Jefferson and Orleans parishes. In 2002, newspapers reported that "the tap through which millions of federal dollars have flowed into southeast Louisiana since 1997 has finally closed."[35] In keeping with the press's dire predictions, the Corps' project manager, Beth Cottone, reported that fifteen of the eighteen separate elements within the SELA project would be disrupted by funding shortages. The preceding year, the Corps fended off interruptions by reallocating funds from other Corps districts. In 2002, federal budget cuts mandated by redistribution of funds into homeland security were behind the anticipated shortfalls at the outset of the hurricane season.[36] The Corps had to suspend work on several components of the SELA project because of funding shortages in 2004. Some of the delayed work was cosmetic landscaping of areas impacted by completed work, but other components involved drainage facilities and pumps that had direct impacts on the removal of excess water.[37] Other delays stemmed from the inability of contractors to meet scheduled deadlines. A pumping station on the West Bank in Jefferson Parish reached completion in October 2004—two years late—and the Corps fined the contractor for not meeting his deadline.[38] But the delay had already occurred.

Stan Green, SELA project manager, reported in December 2004 that he anticipated an improved funding picture for the coming fiscal year. This meant that the Corps would be able to keep existing projects under way but that there would not be sufficient funds to initiate any new work.[39] In both its 2003 and 2004 annual reports, the New Orleans District reported that funding constraints inhibited work on SELA and chose not to report the proportion of work completed.[40]

The SELA drainage improvements yielded mixed results during storms in the early twenty-first century. When Tropical Storm Allison released over twenty inches of rainfall over the course of four days during June 2001, the canals and pumps were able to handle the gradual runoff. Hurricane Isidore in 2002, however, presented a more formidable test. Unleashing up to 23 inches of precipitation over only eight hours, Isidore overwhelmed the drainage system and produced widespread flooding. Even the city's central business district suffered short-lived street flooding. The worst problem occurred when rainfall flooded the interstate highway immediately adjacent to a nearly completed pumping station.[41] This event, with its ironic juxtaposition of flooded highway and incompleted pump, highlighted to all that New Orleans was not prepared for a major hurricane.

PLODDING PROGRESS ON THE LEVEE SYSTEM

Much of the final design for the Lake Pontchartrain and Vicinity Project took place in the late 1980s, and the construction followed during the 1990s. The East Bank projects endured no serious threats comparable to the 1985 flooding caused by Hurricane Juan on the West Bank. Although there was no dramatic impetus to complete the project, most of the East Bank escaped the extensive delays encountered on the West Bank (see chapter 5).

In November 1987, the Corps released its high-level design memorandum for the Jefferson Parish lakefront levee. Two years after the formal decision to shift to the high-level plan, engineers completed their work that would modify the existing levee to meet the revised specifications. Their work would add height to the levees built in the 1940s and augmented by the local levee district with sheet piling in the late 1960s. The older earthen levees combined with sheet piling provided protection between thirteen and fourteen feet (see fig. 4.2b).[42] The new levees would stand seventeen feet high. As part of the larger Lake Pontchartrain project, this component had a favorable benefit-to-cost ratio that recommended construction. The design memorandum called for the lakefront levees to be completed by 2006.[43]

Concerns with preserving wetlands and protecting local scenic waterways in St. Charles Parish prompted a levee realignment well inland from the lake to a position near Airline Highway (U.S. 61). The St. Charles barrier would connect a levee at the western extreme of Jefferson Parish with the downriver guide levee of the Bonnet Carre spillway (see fig. 4.2a). This levee, like its counterpart in Jefferson Parish, would utilize a geotextile material to add strength. Set back a considerable distance from the lake and buffered by an extensive wetland, the design grade for the St. Charles levee was fifteen feet. As with other components of the Lake Pontchartrain and Vicinity Project, Corps planners concluded there was a positive benefit-to-cost ratio.[44] In 2001, the Corps announced that, beyond the levees, crews had completed the Bayou Trepangnier Drainage Structure.[45] It was one of five structures that enabled the Corps to close levee segments where streams intersected the structure's perimeter. Such fixtures remained open until a hurricane threatened. By 2004, the Corps had spent $35 million on the St. Charles project, and the parish contributed another $15 million. Despite the Corps' major progress on the levee, Al Naomi, project manager, reported that several gaps remained and needed prompt attention. With a tight federal budget, the various parishes had to vie for limited appropriations and also employ temporary measures to close the gaps each hurricane season. The 2005 fiscal budget included only $3.9 million for hurricane protection, while the Corps anticipated a need for $22 million in the Lake Pontchartrain region. It estimated that $9 million was required to complete the St. Charles project.[46]

Designs for "remaining work" in New Orleans continued into the early 1990s. In 1993, the Corps released its design memorandum for several components along the Orleans Parish lakefront. Each would bring a portion of the lakefront protection up to the high-level standards. The standard project hurricane for these projects remained a 100 mph storm, although with a return frequency of 300 years and a forward speed of 7 mph. This memorandum sought to raise structures designed and built under the barrier plan and rendered inadequate by subsequent adjustments to the overall plan. High-level standards called for these floodwalls and associated floodgates to stand 3 feet higher than they did in the early 1990s (see fig. 4.2c).[47] Incrementally, the Corps pushed ahead with construction of these components. By 2001, the Corps reported that work was 88 percent complete on the Lake Pontchartrain and Vicinity Project, and, by 2004, progress stood at 90 percent complete.[48]

Included in the Lake Pontchartrain portion of Orleans Parish were efforts to complete the parallel levees along the drainage canals—17th Street, London, and Orleans (see fig. 6.3).[49] These three canals transport runoff lifted from low areas of the city to Lake Pontchartrain. They range in length from 2.5 to 3.1 miles in length and represent an early-twentieth-century drainage system with massive pumps situated at the low points of the city on the river side of the Metairie—Gentilly ridges. From that position, the pumps could lift water into the canals, where it would flow into the lake. The Corps developed two options for protecting areas near the canals: butterfly gates

FIGURE 6.3 Seventeenth Street Canal parallel levees, 2001. Photograph courtesy U.S. Army

at the canal mouths that it could close when a hurricane threatened flooding or raised parallel levees along the canal. With a responsibility to pump runoff that would collect during a major storm, the Sewerage and Water Board opposed the construction of butterfly gates at the mouths of the canals (see chapter 4). It considered the gates an impediment to its responsibilities and feared flooding would result when the canals, unable to empty into Lake Pontchartrain, overflowed and spilled water back into the city streets. The

Sewerage and Water Board's concern had been accentuated with the spate of heavy rains between 1978 and 1995 that repeatedly had produced floods in the city. The Corps continued to push for the butterfly gates, but local interests resisted the decision. In 1988, a Corps planning document clearly recommended the butterfly gates for the Orleans Avenue Canal. A year later, the Lower Mississippi Valley office in its review of the London Avenue Canal design memorandum noted that "local interests must concur with the butterfly control valve plan before you proceed with the recommendations." The report asserted that some considerations placed greater emphasis on optimal operation of the drainage system pumps. It added that the parallel option had become more "desirable" when the Sewerage and Water Board called for establishing pre-agreed conditions for closing the gates. Neither organization wanted the other deciding when to close the gates. A year later, the design memorandum for the 17th Street Canal indicated the Corps had accepted the parallel levee plan despite its initial preference for the gates.[50] Given local concerns, the Corps proceeded with the alternate plan, which consisted of raised parallel levees built sufficiently high to hold back surge and waves from a project storm (with 100 mph winds). The designs called for a combination of earthen levees topped by floodwalls along the canals. Levee heights ranged from 13.5 to 15 feet along these canals. In addition, the plan called for enlarging the capacity of the canal itself. This called for remedial work on the parallel levees. The Corps reported in its design memorandum that the Sewerage and Water Board (which had responsibility for the eastern levee) had been cooperating with the Corps on this, although the East Jefferson Levee District (which had responsibility for maintenance of the west levee) had taken a "wait and see attitude" toward the remedial work.

The Orleans Levee District had also shown its displeasure with the butterfly gate option by initiating design work on the parallel levee option in the 1980s before the Corps had finalized its decision. After initiating work, high cost estimates to complete the work prompted them to postpone construction. This enabled them to spend their limited funds on construction after the Corps approved the parallel levees in 1990 and receive cost-share funds. The Orleans Levee District reported that it favored the Orleans Canal parallel levees because it would have to maintain the existing parallel levees regardless of which option the Corps selected.[51] Ultimately, overwhelming local preference for the parallel levees led to their construction.

Funding issues clouded progress of the loop levee around the populated areas in St. Bernard Parish. When the Corps announced that it would seek $6.7 million in cost-share funds from St. Bernard Parish in 1990, the parish expressed doubts it could meet that obligation. Indeed, the police jury passed a resolution requesting Congress to forgive the parish debt. Parish officials continued to begrudge the Corps for creating, in their view, a flood risk with the construction of MRGO. Perhaps more important, the volume of ship traffic, port activity, manufacturing, and associated economic development projected with the canal never materialized, and costs of the hurricane-protection project multiplied more than eight times the original estimates, increasing the parish's cost-share commitment. The parish enjoyed a brief moment of relief when a congressional committee approved a debt forgiveness proposal. Later that year, Congress stripped the debt waiver from the legislation, and once again the parish faced burdensome costs.[52] Despite uncertainties about the financial capabilities of local partners, work continued in St. Bernard during the 1990s.

Much of the construction done in the late 1990s and early 2000s included closing many small gaps and bringing previously built sections up to the protection level required by the high-level plan. These components exemplified what the Corps referred to as the "remaining projects," as did the various components in St. Charles Parish. In 2001, the Corps initiated work to raise the combined levee/floodwall along the 17th Street Canal as much as three feet. Subsidence since its construction in 1992 had rendered it a vulnerable component. This project consumed some $1.6 million of hurricane protection funds.[53] In addition, local interests had to maintain levee sections that both the Corps and the local organizations had constructed. A levee west of Louis Armstrong Airport, built by the airport authority when it extended its main runway, had settled some four feet by 2003. This created a serious gap in the levee system and exposed most of Jefferson Parish to the threat of flooding. After Hurricane Isidore, the airport authority acknowledged a serious problem existed and set out to bring the levee up to grade.[54] Local environmental conditions that permitted subsidence of heavy-flood-protection structures required constant monitoring and adjustment.

In addition to maintenance and repairs, entirely new components began to compete for federal funding. The West Bank project in St. Charles Parish was not part of the original plan.[55] In the post-9/11 era, as the federal government placed a higher priority on homeland security, competition for funds became more intense. According to Senator Mary Landrieu's office, appropriations for hurricane protection fell substantially during the 2002–5 time span. In 2002, Congress funded the Lake Pontchartrain and Vicinity Hurricane Protection Project with $14 million. That amount fell to $7 million in 2003, to $5.5 million in 2004, and rebounded slightly to $5.7 million in 2005.[56] Tight budgets were a constant impediment to letting new contracts during this period, even if they did not interrupt ongoing projects.[57] Carol Burdine, a project manager for the West Bank and New Orleans to Venice components, indicated that during the years before 2005, funding had been tight and limited the Corps to construction and not additional planning and design.[58] Al Naomi, project manager for the Lake Pontchartrain and Vicinity components, acknowledged that funding had been tight during the several years before Katrina but did not consider the emphasis on homeland security as the problem's source.[59]

Despite tight budgets, beginning in 2002, the Corps sought funds to undertake a study of the feasibility of upgrading the New Orleans area hurricane-protection system from its ability to guard against a category 3 to a category 5 storm. Al Naomi, a senior project manager, stated "if we are going to invest hundreds of millions in systems to evacuate people from the city, maybe we should place that money into a system to keep storm surges out of the city in the first place." Estimates placed the cost of the study at seven to nine million dollars. This initiative recognized the perilous situation faced by coastal Louisiana and the increased risk presented by coastal land loss and subsidence of the delta. In effect, it acknowledged that the benefit-to-cost calculations had changed with the loss of barrier islands and coastal wetlands.[60] The New Orleans District continued its efforts to obtain funds to evaluate a category 5 system the following year. Local officials supported this effort and in public statements equated hurricanes with terrorists to try to place hurricane protection on equal footing with homeland security priorities. Preliminary estimates placed the cost of category 5 protection at more than $2 billion.[61]

In the course of its 2004 inspections of all flood control works in the New Orleans District, the district reported on each component of the hurricane-protection system. Enlargements of the lakefront levees were incomplete in Jefferson Parish. The district engineer reported that maintenance on the existing levees and floodwalls was exceptional and that he anticipated letting contracts on two reaches of the levee enlargement work if funding became available. He reported that work on the floodproofing of the bridge over the 17th Street Canal would continue and that he expected its completion in fiscal year 2005. Overall, the flood control works in the East Jefferson Levee District received an "acceptable" rating.[62]

The district engineer reported that hurricane-protection works in the Orleans Levee district were 90 percent complete, with a completion scheduled for 2009. Outstanding tasks included flood proofing bridges at the 17th and London Avenue canals, fronting protection for the pumping stations, levee enlargements, and floodwall cappings. He assigned an "acceptable" rating for the works in the Orleans Levee District.[63]

Other elements of the system included 43 miles of hurricane levee in the Lake Borgne Levee District that was up to grade and received an "acceptable" rating.[64] In Plaquemines Parish, sixteen miles of hurricane-protection levee in the Grand Prairie Levee District were 90 percent complete and earned an "acceptable" rating. Final lifts were scheduled to begin as soon as funding became available.[65] Plaquemines Parish West Bank Levee District had responsibility for three miles of hurricane protection levee, and its structures received an "acceptable" rating. Also in Plaquemines Parish, the Buras Levee District included 34 miles of hurricane-protection levee. Work remained to raise segments of river levee to hurricane-protection grade. The report stated that "due to funding constraints, the only on going contract in this area at the time of the inspection was the Fort Jackson to Venice second lift levee enlargement project." Levees received an "acceptable" rating.[66]

The South Lafourche Levee District held responsibility for the Larose to Golden Meadow hurricane-protection levees. This project with 43 miles of levee and 6 pumping stations was still under construction in June 2004. It received an "acceptable" rating as part of an interim inspection.[67]

Grand Isle endured damages during Tropical Storm Isidore and Hurricane Lili in 2002 and additional damage due to Tropical Storm Bill in 2003. The Town of Grand Isle had requested funds to effect repairs and to restore the damaged section of artificial dune. This repair work was still under way in August 2004 with an anticipated completion date of December that year.[68]

The Pontchartrain Levee District in St. Charles Parish maintained 10 miles of hurricane-protection levee. As of September 2004, this work was 60 percent complete and "the protection system is not yet closed." The district engineer noted that crews could close the gaps in the system in fiscal year 2005 if funds became available. Project officials indicated a completion date of 2013.[69] Numerous small projects hinged on receiving funding, which were uncertain on the eve of Katrina.

RELATED NONSTRUCTURAL PROJECTS

Evacuation and Dewatering

While planning and preparations for hurricanes depended a great deal on structural means of protecting the urban population, experience proved repeatedly that evacuation was an effective strategy for southeast Louisiana and had been a fundamental part of storm preparations for a century. As work progressed on the post-Besty hurricane-protection system, evacuation never diminished in importance. In fact, it became even more critical as levees encircled the region. Higher levees around a larger territory could hold more overflow and actually increase flood depths, rendering the old local evacuation practices ineffective. Additionally, subsidence within the levee ring further exacerbated the potential for calamity.

Evacuation plans and the public's execution of them have changed over time. When Hurricane Betsy blew on shore in 1965, somewhere between 250,000 and 500,000 people in the lower river parishes evacuated their homes. Without the convenience of interstate highways, evacuation tended to be local. Citizens in low-lying neighborhoods of New Orleans and in St. Bernard and Plaquemines parishes received instructions to relocate to evacuation centers (generally schools or military facilities). Most schools were two- or three-story building and offered greater protection against flooding, plus they had facilities for large numbers of people. Officials considered this the safest option in the 1950s and 1960s. Localized population relocations allowed people to leave the most vulnerable situations. With only minimal levee protection, there was less risk of high-level flooding that would inundate the natural levee. Although some residents remained in low neighborhoods during Betsy, the loss of life was modest when compared to Hurricane Katrina.[70]

Following Betsy, interstate highway construction contributed to urban sprawl and also shifts in evacuation planning. Freeways encouraged longer commutes on a daily basis and encouraged emergency preparedness officials to promote longer-distance evacuations. In a hurricane preparedness study prepared in collaboration with the Federal Emergency Management Agency and the National Weather Service, the Corps developed various hurricane models to forecast vulnerable populations and locations. Within a nine-parish hurricane-exposed area of southeast Louisiana, the Corps identified as many as 1.2 million people (1990 population) that would be vulnerable during storms with an intensity of category 4 or greater.[71] The Corps reported that "up to 15 percent of the residents living in the city of New Orleans are without their own transportation and would rely on public transportation for assistance in evacuating." For a category 4 or greater storm, that would include more than 66,000 people in vulnerable areas of the city. In addition, the multi-agency report noted that surveys conducted after Hurricane Elena indicated southeast Louisiana residents evacuated as if they lived in a low-risk area, and that Vietnamese respondents indicated that they evacuated at about half the rate of other residents of the area. These were troubling numbers to emergency preparedness officials, but the report offered no solutions. And personal vehicle transportation via interstate highways became the dominant element of evacuation planning.[72]

Shortly after this 1994 report, Hurricane Georges blew on shore just east of New Orleans. A stern warning from the National Weather Service prompted a massive evacuation that produced giant traffic snarls on the freeways. As tens of thousands took to the highways, drivers found that it took six hours to flee the eighty miles to Baton Rouge.[73] When the storm veered eastward, New Orleans dodged a direct hit, but Georges provided a dress rehearsal for mass evacuation via the limited interstate highway system and produced disappointing results.

In 2000, the Corps prepared a plan to remove flood water following a hurricane that overwhelmed the levee system. The report pointed out that there were thirteen separate protected loops that would be susceptible to flooding from a category 3 or greater storm. A category 5 storm, it noted, could overtop all the levees and produce widespread flooding. Rainfall and the possibility of pump failure could add to accumulations of water within the individual "bowls." If storm surge and waves overtopped the levees, water could conceivably rise to the lip of the levee system. In the event of levee failure, some flood water would drain out after the storm surge subsided. This report acknowledged the stark reality that a massive storm could overwhelm the city, and the document represented an element of planning for such an eventuality.[74]

In 2001, the Corps released a preliminary transportation model as a preface to a more complete evacuation study. It noted that southeast Louisiana "is one of the most vulnerable to hurricanes in the entire country." The model sought to provide nonspecialist local officials with a tool to estimate times necessary for parish populations to clear the vulnerable areas.[75]

After the passage of Hurricane Georges (1998) and publication of the Corps' studies on hurricane preparation, there were numerous proclamations about the serious risk that a major storm posed to New Orleans. FEMA, national science magazines, and the *New York Times* all reported on the region's susceptibility.[76] Local residents who were not attuned to these sources were not spared the warnings. The *New Orleans Times Picayune* ran a major series pointing out hurricane risk in no uncertain terms. The first article opened with this unambiguous and prescient passage: "A major hurricane could decimate the region, but flooding from even a moderate storm could kill thousands. It's just a matter of time."[77] Corps officials participated in the discussion. Project managers from the two major components regularly gave presentations to civic groups underscoring the situation that New Orleans faced.[78] Scientists at Louisiana State University shared their hurricane flooding predictions with the press and posted models that depicted flooding on the Internet.[79] Scientists at the University of New Orleans and the state geological survey also participated in studies of evacuation and coastal land loss that added to knowledge about the city's vulnerability.[80] A host of public officials and media shared in the task of warning local populations about the risk of a major hurricane strike. These numerous sources abundantly document the critical nature of the situation, and experts, public officials, and the general public all shared awareness of their precarious situation, even if only in general terms.

As Hurricane Ivan bore down on Louisiana in 2004, officials urged citizens to evacuate, and area residents hit the highways once again. Despite what turned out to be a false warning when Ivan veered eastward and spared the New Orleans area, the evacuation prompted a massive response. Some 600,000 people swarmed onto the limited outbound routes. Travel times to Baton Rouge (eighty miles away) exceeded eight hours for some. Excessive congestion caused concern among emergency planners who feared the unpleasant experiences associated with Ivan might deter evacuation when the next storm bore down on southeast Louisiana. Emergency planners put in place a "contra flow" plan for future evacuations that would direct traffic on all interstate highway lanes to move outbound and thereby facilitate a larger number of evacuees to flee the incoming storm.

Coastal Restoration

Scientists began reporting on the coastal land loss issue in south Louisiana in the early 1970s. Subsequently, numerous investigations sought to monitor and analyze the various aspects of wetland erosion and subsidence. By the late 1980s, there was abundant documentation of land loss associated with sediment starvation, subsidence, erosion, canalization of the wetlands, faulting, and sea level rise. With compelling scientific evidence in hand, Louisiana embarked on a campaign to counter the coastal land loss process. In 1990, its congressional delegation, led by Senator John Breaux, convinced Congress to pass the Coastal Wetlands Planning, Protection, and Restoration Act (P.L. 101–646). This act provided federal funds, to be matched by a Louisiana contribution, to underwrite various priority coastal restoration projects.[81] Several state and federal agencies have partnered in this massive effort that included a considerable contribution from the Corps of Engineers. Although not explicitly a hurricane-protection project, healthy wetlands and barrier islands can serve as a buffer against storm surge and waves and thus contribute to hurricane protection.

Even before the so-called Breaux Act, the Corps undertook a study of coastline restoration projects.[82] In addition, projects that impact coastal wetland restoration built with considerable Corps participation began taking shape in the 1980s. One of the first to reach completion was the Caernarvon Freshwater Diversion Project. Operational since 1991, the Caernarvon diversion consists of gates set in the river levee downstream from New Orleans. When opened, the gates allow freshwater that contains nutrients to flow into Breton Sound. The primary purpose of this project is to sustain the wildlife and fish habitats. It also serves to deliver sediment that gradually offsets land loss owing to subsidence and erosion. A second and larger project, Davis Pond Freshwater Diversion Project, opened in 2002. It sends nutrients and sediment from the Mississippi River upstream from New Orleans into the upper Barataria Estuary.[83] In 2002, the Corps also completed the West Bay Sediment Diversion project in Plaquemines Parish. This project involved cutting a gap in the natural levee in the bird's foot delta natural levee to allow the river to deposit sediments in the area adjacent to the main channel. The purpose is to introduce sediment and establish vegetation, thereby restoring a portion of the delta lost to subsidence.[84] The Corps has also participated in projects to rebuild and stabilize the barrier islands, other than Grand Isle, off the Louisiana coast. Dredges lift sand from offshore shoals and place it on the barrier islands. Other efforts to protect the barrier islands include replanting dune vegetation and placement of longshore erosion control structures. Joint projects have enhanced some 75,000 acres of barrier island land.[85]

Requests for an additional $14.5 billion in federal funding to underwrite the coastal restoration effort met with resistance in Congress before Katrina. Coastal restoration is an extremely slow process, and effective efforts will take decades. Such was also the case with the levee system. Shoring up the state's wetlands is vital to combat the effects of sea-level rise and to dampen storms before they strike the levees that span much of southeast Louisiana.

■ ■ ■ ■ ■ ■ ■

CONCLUSIONS

In the years since Hurricane Katrina roared ashore, individuals and teams have authored countless accounts of the storm and its impact. Numerous quickly written books,[1] assessments of the levee failure,[2] and government investigations have reviewed the dramatic event from different perspectives.[3] Those works focused largely on what happened during and immediately after the category 3 storm that became a catastrophe on an unrivaled scale. With different purposes and distinct conclusions, these chronicles made the first attempts at assigning fault for the levee failures, the ineffective social and political responses, and the challenges that face rebuilding a region. Such accounts are essential in understanding the brutal results of Hurricane Katrina.

This book has taken a different tact. Its purpose has been to provide a longer-term perspective of the processes that led to the construction of the hurricane defense system in the New Orleans vicinity, and in no way does it attempt to evaluate the engineering or the structures themselves. It considers the initiation of federal involvement in hurricane preparations in the 1940s and concludes before Katrina rent asunder more than fifty years of preparations. A historical treatment adds perspective to the post-disaster assessments. It provides a richer context by tracing the multiple strands of engineering, political maneuvering, financial challenges, legal wranglings, and public participation. And it does not have to focus on the elements that failed. Indeed, there were failures that Katrina did not reveal.

It exposes the human element woven into the fabric of the levees and drainage systems. After all, the levees are not just structures. Many individuals contributed to their design and construction, to securing funding and cooperative pacts among different units of government, to the recommendations for adjustments to the system, and, most important, to the decisions to live within their perimeter. The human dimension of Hurricane Katrina was all too obvious in the media coverage of the storm's aftermath, but it received too little consideration in the preparations. Science and engineering guided preparations of the levees, drainage works, the potential impacts of the storms, and even evacuation. The disaster of Katrina was not the meteorological event but the inadequate preparations for the challenges faced by the humans who opted to live in a hazardous location and the choices they faced as a major storm headed their way. In this book, I seek to illustrate the complexity of one component of the human preparations for powerful weather events: the hurricane-protection system. There is no easy way to prepare a benefit-to-cost analysis or a three-dimensional model that adequately measures human endeavors in this process. There are no formulae that precisely predict the relationship between the damages of the last storm and preparations for the next. I do not seek to trace all the myriad threads of this story, but I hope this book reveals a small part of the infinitely complex web of human activities that resulted in a hurricane-protection system around portions of New Orleans and southeast Louisiana.

LONG-TERM TRENDS

The Perpetual Process

Several things stand out in reviewing the nearly sixty-year involvement of the Corps of Engineers in hurricane-protection efforts in the New Orleans area. The most ironic is the fact that major Corps initiatives to improve hurricane protection preceded major storms. In 1945, local organizations appeared before the Corps' Board of Engineers and made an appeal for improved protection along the Lake Pontchartrain shore. The Corps responded with a recommendation for improved protection that same year, and Congress authorized work on such a project in 1946.[4] The unnamed hurricane of 1947 dramatically emphasized the need for lakefront protection and prompted congressional authorization of an expanded phase of Corps involvement.

In addition, the Corps completed an interim survey report of hurricane-protection needs in November 1962. By June 1965, they advanced to Congress a plan for a comprehensive hurricane-protection system for the Lake Pontchartrain and lower delta region. This plan was before Congress when Hurricane Betsy struck in September 1965.[5] When Congress appropriated funds to proceed with the Lake Pontchartrain and Vicinity Project, the Corps was able to respond quickly at the outset.

A series of news and scientific publications had highlighted New Orleans's vulnerability and produced heightened public concern about the "big one." In 1990, Bob Sheets, director of the National Hurricane Center, warned emergency management officials that a major hurricane strike at New Orleans would kill more people than a similar event in any other U.S. city.[6] Local journalists and national science writers offered similar projections on the inevitable calamity.[7] In 2002, the Corps of Engineers sought funding to consider the costs and feasibility of augmenting the New Orleans area defenses to withstand a category 5 storm.[8] Since Katrina, Congress approved funding for an investigation of building category 5 protection, and the Corps has released a preliminary technical report.[9] Any additional fortification of the existing system, such as upgrading it to category 5 level of protection, will add substantial time to the construction of hurricane protection. And already, the technical report appeared behind schedule.

The irony of these requests for improved protection systems lies in the fact that, while timely, even with immediate approval and funding, the Corps' contractors could not have completed construction in time to prevent damages by the storms that followed each solicitation. Nonetheless, the appeals for improved protection indicate that local residents and officials and the Corps of Engineers recognized a continually expanding need for enlargements to the existing system, and that concern was not dormant. As large numbers of suburban residents moved into previously thinly settled areas, they sought security from tropical storms. As development swept across more and more subsiding wetlands, expectations for more effective protection rose. Constant pressures to enlarge and improve the hurricane protection system ultimately contributed to the costs and protracted construction phase. Hurricane-protection provision became a perpetual process.

Environmental Challenges and Uncertainty

The deltaic environment upon which the Corps built much of the hurricane-protection works is highly dynamic. Building levees along the river altered one of the key elements of this system long before hurricane-protection planning began. By funneling spring floods between river levees, local and later federal levee builders sent millions of tons of sediment into the deep gulf that once added mass to the delta. Sinking under its own weight, the region is subsiding. Canals cut for oil exploration and the removal of subterranean oil and gas deposits have contributed to coastal wetland erosion and regional subsidence. Furthermore, building ring levees and draining the peaty soils within their perimeters contributes to the lowering of the land level. Add rising sea levels to this picture, and it becomes apparent that foundation conditions for weighty levees are far from ideal, and prognoses for levees providing long-term protection at their constructed height become highly problematic.

Corps engineers took subsidence into account in their initial plans and sea level rise in later components as they designed the hurricane-protection system. Its engineers recognized that multiple levee lifts were a fundamental part of constructing a viable hurricane-protection defense system in southeast Louisiana. The environmental conditions contributed to a lengthening of the construction schedule—a situation known to the designing engineers. The Corps, while not responsible for initiating levee building along the lower river, has played a major role in the construction and operation of these structures for over a century. It recognizes the impacts of sediment starvation and has embarked on projects in collaboration with local interests to restore the fragile delta. Such projects are slow and produce incremental results.

Debates over environmental impacts forced alterations in the plans and contributed to delays. Local environmental groups and commercial fishermen challenged the Corps' assertion that the initial barrier plan would not modify conditions in Lake Pontchartrain. Local business interests feared the barrier plan would divert surge across St. Tammany Parish and also challenged the initial plan. When the federal court blocked the barrier plan in 1978, it did not reach its conclusion based on conclusive evidence that the barrier would damage Lake Pontchartrain; rather it claimed the Corps had presented "inadequate" science in its environmental impact statement.[10] Complex conditions rendered scientific analysis uncertain to some. That uncertainty produced a court decision that forced engineers to reconsider the barrier plan and ultimately opt to begin the long process of planning for the high-level option. With its larger levees, this plan required more levee lifts and more time, straining environmental limits of the region's structural foundation.

Local-Federal Tensions

Tensions between federal and local governments have deep roots in New Orleans and Louisiana. When Louisiana became a U.S. territory in 1803, some Creoles resented the upstart Americans who arrived in their city. Legal conflicts over control of the batture along the riverfront and questions about the legality of charging shippers a local levee tax stand as clear examples.[11] Distrust of the Corps surfaced in the lawsuit that alleged the Mississippi River Gulf Outlet caused flooding during Hurricane Betsy in 1965. Other examples, such as challenges to different lock proposals and even battles between citizens and the U.S. Environmental Protection Agency over the resolution of the Agriculture Street Landfill, reveal that the long-standing unease between local and federal interests continues to the present.[12]

There are numerous examples of this tension that contributed to the final form of the hurricane-protection system. Some led to minor adjustments with minimal delay, others hampered progress, and others compromised the level of protection. From the initial announcement of the Corps' hurricane protection plan in 1962, St. Bernard Parish residents sought to modify the plan. A key thrust of their requests was the enlargement of the territory to be encompassed by the levees. Adjustments in the post-Betsy plan expanded the land area protected, although not to the full extent that some desired. This discussion did not prove to be exceptionally contentious, but eventually it strained the parish's relationship with federal bodies. Over time, the adjustment added to the parish's cost. When parish leaders appealed to Congress to forgive its growing debt, Congress refused, exposing a frustration at the federal level to satisfy incessant local requests.

The most pronounced delay prompted by federal-local disagreements involved the West Jefferson hurricane-protection project. Parish officials expressed dissatisfaction with the limited area the Corps planned to encircle with its levees. In 1981, they opted to take over the levee-building project in order to expand local development potential. In 1984, the Corps denied the parish's levee alignment application, and the parish council voted to return the project to federal authorities. This delay extended the construction schedule of West Bank protection.

The Sewerage and Water Board of New Orleans (S&W Board) questioned the Corps' proposed butterfly floodgates at the mouths of the three outfall canals in Orleans Parish. Conflict over this issue stemmed from competing responsibilities. The Sewerage and Water Board had responsibility for expelling runoff during heavy rainfall. Since hurricanes generally deliver copious amounts of precipitation, such events make operation of the city's pumps during the storm vital to keeping up with drainage demands. The Corps had responsibility for preventing hurricane damage to the lakefront areas traversed by the outfall canals. Constructing butterfly floodgates could have reduced the costs of levees along the canals and provided a single line of protection along the lakeshore. Local interests won this debate, and parallel levees along the canals substituted for butterfly floodgates.

Sense of Urgency

After Hurricane Betsy, local politicians and other public authorities all shared a heightened concern with the need for hurricane protection across southeast Louisiana. Calls for federal assistance came from all quarters, and Congress promptly authorized funding. The Corps had a plan before Congress in 1965, and this helped jump-start the huge job of designing and building the multifaceted hurricane-protection system. As progress bogged down and construction fell behind schedule, Congress investigated and chastised the Corps for the project's cost increases and its failure to stay on schedule. Local organizations frequently voiced concern about the pressing need for the project to move forward expeditiously.

Local organizations, legal challenges, sitting administrations, and other contingencies offset repeated appeals for efficiency and contributed to substantial alterations in the plan and subsequent delays. Despite repeated warnings of the impending "big one," urgency eroded. Levee construction in the wetlands of southeast Louisiana requires repeated phases of construction, which demand years to complete. As the Corps made deliberate construction progress, years turned into decades. Several close calls produced short-lived renewals of urgency, but there was no quick fix, and the Corps, local officials, and residents had to await the gradual completion of the protection works. All the while, expanding suburbs demanded new levees to protect areas that did not need protection at the outset of planning.

Hurricane Katrina was not the worst-case scenario, which forecast no levee failure and massive flooding owing to levee overtopping. Such circumstances could have produced even greater flooding and inundations that required greater pumping to expel water from within an intact levee system. The several levee breaches produced during Katrina allowed water in but also allowed some of the floodwater to drain back out. While not the calamity some had predicted, Katrina exposed flaws in the protection system and all other aspects of emergency preparedness. It also instilled a renewed sense of urgency that likely exceeds the one inspired by Hurricane Betsy.

POST-KATRINA CONCERNS

While much can be, and has been, written about the rebuilding of the New Orleans, in this section I will briefly update several elements of the Corps' hurricane-protection works. These include repairs to the existing levee system, the investigation of category 5 levees, the revised procedure for assessing risk, the deactivation of the Mississippi River—Gulf Outlet, and the suit brought against the Corps stemming from the levee failures. These developments have direct ties to the pre-Katrina effort to erect a hurricane-protection system. While the city's recovery is vitally important, it is beyond the scope of this work. Nonetheless, restoring structural protections and ensuring a basic level of safety underlies all rebuilding efforts.

Even after Katrina moved ashore and as it whipped inland through Mississippi, levee repairs began. Dramatic news images showed helicopters dropping massive sandbags to fill the breaches along the ruptured 17th Street Canal. As quickly as the city was drained (fifty-three days), Corps engineers and contractors set to work restoring the levees to their prestorm designed strength. Crews delivered soil, and heavy equipment operators shaped the earthen levees adjacent to the MRGO and the Intercoastal Waterway. Along the outlet canals and the Industrial Canal floodwall, workers installed temporary coffer dams that enabled the reconstruction of damaged sections. The initial intent was to repair the floodwalls and have basic repairs in place by the June 1, 2006, hurricane season. The Corps set September 2007 as its target for raising undamaged components of the system to their designed height. Certification of the entire system—that is, completion of the entire system up to design grade—has a 2010 completion deadline.[13] This schedule outpaces the prestorm timeline and underscores the high priority given this effort. Obviously, the storm compelled all players to accelerate what had become before 2005 a perpetual project.

Post-Katrina repairs include improvements for several components. Plans called for armoring certain sections of earthen levees with stone riprap. This effort will reduce erosion caused by water overtopping the levees and reduce the risk of levee failure.[14] For floodwalls with I-wall construction that failed during Katrina, replacement sections have stronger T-wall or L-wall construction (anchoring structures that provide additional bracing) and have less "stick-up" floodwall standing above the earthen levee to reduce pressure from surge. Corps plans also called for floodproofing pumping stations to enable them to continue operating in the event floodwaters enter the city in the future.[15]

In addition, the Corps received approval to install gates at mouths of the outfall canals in Orleans Parish. To offset prestorm concern about flooding produced by water backing up in the canals, the Corps installed temporary pumps to redirect water around the closed gates.[16] Also scheduled for completion at the beginning of the 2006 hurricane season, the Corps did not make that deadline but did have the temporary pumps in place by late summer 2006. These pumps do not have the same pumping capacity as the pumps that feed the canals, but larger, permanent pumps had been installed by 2008.

Perhaps the most impressive element of the post-Katrina flood protection is the addition of a new surge-protection project to be built at the confluence of the Intercoastal Canal and the MRGO. Storm-surge models revealed a "funnel" effect produced by the converging levees that paralleled those two waterways. The surge-protection project is to armor eastern New Orleans, St. Bernard, and the 9th Ward from the type surge produced by the combined effect of hurricane winds, low pressure, and the confinement of high water between levees. Officials had not projected a completion date by summer 2008.[17]

Well before Katrina, the Corps had prepared a preliminary reconnaissance study of strengthening southeast Louisiana's hurricane defenses, including the option of providing protection against category 4 or 5 storms. In the course of its initial study, the Corps concluded a full-fledged feasibility study would cost another $8 million.[18] Katrina spurred Congress to appropriate funds to begin that feasibility study, which was due by the end of 2007. The Corps' explicit mission was "to conduct a comprehensive hurricane protection analysis and design; to develop a full range of flood control, coastal restoration, and hurricane measures exclusive of normal policy consideration for South Louisiana; and to submit a final technical report for 'Category 5' protection within 24 months."[19] It did not achieve its 2007 deadline but issued a "Technical Report" in February 2008. This report steers away from discussing category 5 protection. Rather, hydrologists created models of flooding for a 100-year event, a 400-year event, and a 1000-year event (a 100-year event represents a 1 percent chance of occurrence each year). They used these to assess potential surge and its impact on local development. Since Katrina, planning has centered not on providing a greater level of protection but on bringing protection up to the 100-year storm level. The technical report considers category 5 storms, but this is not the focus of the assessment that has fallen behind schedule. On the positive side, the Corps' technical report supports the "multiple lines of defense" approach, which relies less on a single barrier but seeks to employ numerous integrated defenses, including nonstructural options. Indeed, the report emphasizes the risk of calamitous outcomes associated with the failure of structural defenses.

Initial plans for structural defenses were based on the standard project hurricane, a measure that changed over time but one that emphasized wind speed, storm track, and forward motion. Different sections of the structural defenses had different SPH and consequently different design criteria for structures. The Corps' Risk and Reliability study conducted since Hurricane Katrina seeks to emphasize not the storm category (which was not a measure of importance in the initial planning either), but storm size and intensity. Planning in the post-Katrina era relies on storm probability (not the hurricane category) to provide protection against a 100-year storm (1 percent chance annually). By providing protection against a 100-year storm, the levees will enable mapping the 100-year floodplain within the levees based solely on rainfall events and not hurricane surge. Yet, moving away from a wind-speed measure of risk undermines public comprehension of risk. For decades, weather forecasters have been able to capture the attention of the public with warnings of category 4 or 5 storms. It will take a while to establish a clear message using storm size and intensity.[20]

Since Hurricane Betsy in 1965, St. Bernard Parish residents have charged that the MRGO contributes to hurricane-caused flooding. Since Katrina, this issue has regained prominence and has contributed to the congressional authorization to close the MRGO. Although the courts rejected a post-Betsy lawsuit that charged the Corps with responsibility for flooding (see chapter 4), public apprehension lingered. Also, during the forty years between Betsy and Katrina, the MRGO never developed the level of traffic projected and was not the economic benefit forecast. This undermined public support for it. Erosion of the banks of the canal and its contribution to the overall wetland loss made MRGO the object of a strong campaign to close it. When Katrina pumped tons of sediment into the canal and reduced its depth, opponents saw an opportunity to press for closure.

Responding to local interests, Congress provided funding for the Corps to draft recommendations and an environmental impact statement for closing MRGO.[21] The plans call for constructing a stone barrier near the mouth of the canal and integrating this project with the surge-protection barrier. This eliminates the need to redredge the Katrina-laid sediment and also closes the channel to commercial shipping. The decision to deactivate MRGO became official in June 2008 and thus ends an acrimonious chapter is the Corps' involvement with St. Bernard Parish.

Yet, it does not end public claims that the Corps' works contributed to flooding. A series of post-Katrina lawsuits charged that the levee failures in Orleans Parish made the federal government responsible for damages sustained by residents. Individuals and businesses leveled nearly 500,000 claims with monetary damages that exceeded $3 quadrillion. Consolidated as one large suit, a federal judge ruled that the Flood Control Act of 1928 provided immunity for the federal government from liability stemming from the failure of flood control works. While ruling in favor of the federal government, the judge was highly critical of the Corps and Congress for its protracted hurricane-protection efforts. He wrote that "millions of dollars were squandered in building a levee system with respect to these outfall canals [17th and London Avenue] which was known to be inadequate by the Corps' own calculations. The byzantine funding and appropriation methods for this undertaking were in large part a cause of its failure."[22]

A separate suit remained before the courts during the summer of 2008. The complaint alleged that levee failures in St. Bernard and eastern Orleans parishes stemmed from MRGO. In May 2008, a federal judge concluded that the suit *Tommaseo, et al. v. the United States* should go to trial. By claiming that the MRGO, a navigation work, produced the flooding, rather than the failure of flood-protection structures, plaintiffs in this case avoid the immunity clause of the 1928 Flood Control Act.[23] This lawsuit, regardless of the outcome, will not interfere with plans to deactivate the MRGO, but it may open the door to additional costs for the federal government in this troubled region.

Locally, criticism of the Corps has run deep since Hurricane Katrina battered its defensive barriers around low-lying areas of southeast Louisiana. But it is important to recall that multiple players influenced the timing and placement of parts of the hurricane-protection system. Those outside the region have questioned the sensibility of living and rebuilding in such a precarious place. But risk has never precluded human settlement. Cities elsewhere thrive on fault lines, drought-prone locations, and in the shadow of active volcanoes. The decision to live in New Orleans is more typical than lunatic. The combination of a flawed flood protection, which in many respects exacerbates the risk from extreme storms, and susceptibility to hurricanes is what sets New Orleans apart. The gargantuan project that began in 1965 continues in the wake of Katrina. With its intensity of purpose renewed, hurricane protection is moving forward, but it remains burdened with the same set of complexities that undermined the previous effort.

EPILOGUE

Two hurricanes impacted the Louisiana cost in September 2008: Gustav on September 1 and Ike on September 12. Gustav enabled the Corps and local organizations to gauge the integrity of the rebuilt levees and related structures, and both hurricanes tested Louisiana's post-Katrina emergency preparations.

As Gustav raked Cuba with category 5 winds in late August 2008, I regularly checked its forecast path while attending a conference in France. As the date of my return approached, the National Hurricane Center updated its projections to indicate a Louisiana landfall a day ahead of its earliest projections. What had been a one-day cushion to arrive in Baton Rouge and prepare for a hurricane had been reduced to a few hours, assuming my flight would be allowed into the local airport. As I anxiously watched the weather reports, New Orleans successfully initiated its city-assisted evacuation for those without personal transport and with special needs and closed its airport. The mayor proclaimed that Gustav would be the "storm of the century." While perhaps an exaggeration, given wind speeds at landfall, the statement impelled citizens to flee to safer locations. Evacuation of those who needed assistance proceeded effectively, and most with personal transportation also fled before landfall.

I touched down in Baton Rouge on the last flight into the airport on August 31 as rain and winds washed over the city. I returned home with our regular New Orleans evacuee house guest for a quiet evening, unsure what lay ahead. During most of the day on September 1, winds howled, live-oak limbs crashed into the yard, and rain pelted the roof, and by early afternoon the combination of events had knocked out electrical power. We hunkered down to await word of what happened in New Orleans.

As we eventually learned, Gustav came ashore west of Grand Isle on the morning of September 1 and followed a northwesterly track across the state. It unleashed sustained winds of about 115 miles per hour (category 3) as it moved over the coastal marshes, and, by midmorning, wind speeds declined to about 110 mph (category 2). To the northeast, New Orleans experienced heavy winds (peak gusts of 56 mph), but none of hurricane strength and, according to the U.S. Weather Service, only modest precipitation (almost six inches in two days). Higher wind speeds buffeted Shell Beach and locations along the lower delta. These winds pushed considerable surge across Lake Borgne into the funnel, where television news crews broadcast minor overtopping along the western flank of the Industrial Canal floodwall. The U.S. Geological Survey's preliminary measurements indicated surge in the

Industrial Canal was about ten feet and about five feet at the lakefront near the 17th Street Canal. In New Orleans, the levees and other structures held, and the new pumps at the outfall canals adequately expelled runoff into the lake. While this storm tested the post-Katrina repairs, it did not provide a full-fledged challenge and certainly not one on the scale of Katrina.

A dangerous surge overwhelmed Grand Isle and the lower delta. At Yscloskey, surge rose to eight feet and surge along the coast near the eye ranged from ten to fourteen feet. The combination of surge and wind-driven waves obliterated large sections of the Corps-built barrier dune at Grand Isle, as expected. This left much of the beachfront fully exposed after the storm passed and prompted some residents to call for more permanent barriers. Erosion to restored barrier islands near the storm's eye at Isle Dernieres was less than anticipated by specialists with the U.S. Geological Survey. A local levee failed on the lower delta, but the federal levees held firm. Thus in the face of a serious storm, the patched-up hurricane-protection structures withstood the test, albeit only Grand Isle's protection structures experienced what might be termed a direct hit.

Only days after Gustav, Hurricane Ike angled across the Gulf on a northwesterly track south of the Louisiana coast. In the early hours of September 12, it made landfall near Galveston, Texas, with sustained winds around 100 mph. While it did not threaten the New Orleans and lower delta area directly, its particular course and sustained winds for several days forced considerable surge across the entire Louisiana coastal zone. Surge reached twelve feet at Yscloskey (more than the Gustav surge!) and approached eight feet in New Orleans's Industrial Canal. The level of Lake Pontchartrain rose several feet and produced flooding on the north shore. Gusts in New Orleans approached 60 mph, although rainfall from the storm was less than two inches. Surge across the unprotected southern marshes was considerable, slow to drain, and flooded several communities with no levees. Grand Isle, once again, saw surge and wave feet sweep over the island and destroy more of the artificial barrier dune and produce serious erosion. Nonetheless, the impact on the southeast Louisiana protection structures was not catastrophic. Storm damage was most pronounced in southwest Louisiana, where no hurricane protection structures exist, and wind speeds and surge were more substantial as the storm approached the nearby Texas shore.

The September 2008 hurricanes undoubtedly will help maintain a sense of urgency in completing the structural protections near New Orleans, and Louisiana officials are pleading for funding toward that end. Yet, the financial storms that swept over the American economy that same month may divert attention and funds toward other priorities. Ultimately, coastal Louisiana remains vulnerable, unable to defend itself, reliant on federal assistance, and subject to multiple social, economic, and political currents that are arguably more byzantine than the highly complex and powerful storms that make it a perilous place to live.

■ ■ ■ ■

NOTES

CHAPTER 1. INTRODUCTION

1. U.S. Army Corps of Engineers, *History of Hurricane Occurrences along Coastal Louisiana*, and U.S. Army Corps of Engineers, *Hurricane Study*.

2. Shallat, *Structures in the Stream*, and White, *Organic Machine*. Since the passage of the 1969 National Environmental Protection Act, environmental advocates have leveled criticism at the Corps for letting engineering priorities take precedence over environmental concerns. An excellent account of an encounter between the Corps and environmental groups is Stine, *Mixing the Waters*.

3. Elliott, *Improvement of the Lower Mississippi River*, 164–70; Pabis, "Subduing Nature through Engineering"; and Reuss, "Andrew A. Humpreys."

4. Cowdrey, *Lands' End*.

5. Arnold, *Evolution of the 1936 Flood Control Act*; White et al., *Changes in Urban Occupance of Flood Plains*; Camillo and Pearcy, *Upon Their Shoulders*; and Barry, *Rising Tide*. Also O'Neil, *Rivers by Design*.

6. Cowdrey, *Land's End*, see chs. 1 and 2, and Colten, *Unnatural Metropolis*, ch. 1.

7. White, *Changes in Urban Occupance of Flood Plains*, 228.

8. Burby, "Hurricane."

9. Steinberg, *Acts of God*.

10. Davis, *Ecology of Fear*; Orsi, *Hazardous Metropolis*. See also, Gumprecht, *Los Angeles River*.

11. Kelman, *A River and Its City*, and Colten, *Unnatural Metropolis*.

12. Castonguay, "The Production of Flood as Natural Catastrophe." Also Tobin and Montz, *Natural Hazards*, esp. ch. 4.

13. Geomorphologists define a floodplain as a land surface created by sediments deposited when streams escape their channels. Under the National Flood Insurance Act, the legal definition of the "100-year floodplain" is the areas susceptible to a 1 in 100 chance

of flooding each year. Since levees that offer at least 100-year flood protection surround New Orleans and its immediate suburbs, the legal "floodplain" does not correspond to the geomorphic floodplain. Nonetheless, the pre-Katrina legal floodplain included much of the city, because of its susceptibility to flooding from heavy downpours that can exceed the ability of the city's drainage system to lift the water out of the levee system.

14. Burby, "Hurricane Katrina."

15. U.S. Department of Commerce, *Model Hurricane Plan for a Coastal Community.*

16. Meyer-Arendt, "Historical Coastal Environmental Changes"; Perez, *Winds of Change*; Bixel and Turner, *Galveston and the 1900 Storm.* See also Dunn and Miller, *Atlantic Hurricanes.*

17. Cowdrey, *Land's End*, 77–80.

18. Steinberg, *Acts of God*, and Burby, "Hurricane Katrina." See also Lewis, *New Orleans*, chs. 3 and 5.

19. Sands, personal interview; Fischetti, "Drowning New Orleans"; and McQuaid and Schleifstein, "Washing Away."

CHAPTER 2. CITY AT RISK

1. Saucier, *Recent Geomorphic History.* Native people also built high ground using shell middens that they created as habitable locations. See Kidder, "Making the City Inevitable."

2. *Ibid.*

3. Colten, "Bayou St. John."

4. Saucier, *Recent Geomorphic History*, 58.

5. Lewis, *New Orleans*, 17.

6. See Cowdrey, *Land's End*, 1–15, and Colten, *Unnatural Metropolis*, 16–46.

7. Elliott, *Improvement of the Lower Mississippi River*; Owens, "Holding Back the Waters"; Harrison, *Swamp Land Reclamation.*

8. Owens, "Holding Back the Waters," 158; Harrison, *Swamp Land Reclamation*, 3; and "An Act Relative to Roads, Levees, and the Police of Cattle," 312–34.

9. Monette, "Mississippi Floods," 427–76, esp. 444 and 461.

10. Colten, *Unnatural Metropolis*, 28.

11. Harrison, *Swamp Land Reclamation*; and Davis, "Historical Perspective on Crevasses, Levees, and the Mississippi River."

12. Elliott, *Improvement of the Lower Mississippi*, 162–63; and Harrison, *Alluvial Empire* I-146 to I-149.

13. "Backwater: Rear of City Inundated"; "Inundation"; "Submerged Suburbs."

14. U.S. Army Corps of Engineers, *Hurricane Study*, 17–21. See also *New Orleans Daily Picayune* coverage of the hurricanes September 30–October 3, 1905; September 17–26, 1909; and September 26–October 9, 1915.

15. U.S. Army Corps of Engineers, *Hurricane Study*, 23.

16. U.S. Congress, House of Representatives, *Lake Pontchartrain, Louisiana*; and U.S. Congress, Senate, *Lake Pontchartrain, Louisiana.* Also see Burk and Associates, *East Bank Master Drainage Plan*, 5.

17. U.S. Army Corps of Engineers, *Lake Pontchartrain and Vicinity Hurricane Protection Project: Final Environmental Impact Statement*, 7–8.

18. U.S. Army Corps of Engineers, *Hurricane Study*, 33.

19. U.S. Army Corps of Engineers, *Hurricane Flossy*, 19–20.

20. U.S. Congress, House of Representatives, *Lake Pontchartrain, Louisiana*, and U.S. Congress, Senate, *Lake Pontchartrain, Louisiana.* See also U.S. Army Corps of Engineers, *Hurricane Flossy*, 5–14.

21. Nineteenth-century drainage is examined in greater detail in Colten, *Unnatural Metropolis*, 32–46. See also, Hardee, *Topographical and Drainage Map of New Orleans.*

22. The creation of the twentieth-century drainage system is covered in Colten, *Unnatural Metropolis*, 83–104. Details of the project appear in Sewerage and Water Board of New Orleans, *Semi-Annual Reports.*

CHAPTER 3. SEASON OF THE STORMS

1. Dunn and Miller, *Atlantic Hurricanes*, 55–57.

2. Elsner and Kara, *Hurricanes of the North Atlantic*, 157. They classified category 3 and higher as "major storms."

3. Another Hurricane Barbara moved along the eastern seaboard in 1953, but is unrelated to the 1954 storm that struck Louisiana.

4. U.S. Army Corps of Engineers, *Hurricane Study*, 34–35.

5. U.S. Army Corps of Engineers, *Public Hearing.*

6. "Orleans May Escape Hurricane's Brunt."

7. "Evacuees Returning Home as Flood Problem Created by Flossy Improves."

8. "Lower Parishes Pushing Relief."

9. "Study of Flood Problem Begins"; and U.S. Army Corps of Engineers, *Hurricane Flossy*, 5–9.

10. U.S. Army Corps of Engineers, *Hurricane Study*, 36.

11. U.S. Army Corps of Engineers, *Interim Survey Report*, 1, and Public Law 71, 1955.

12. U.S. Army Corps of Engineers, *Hurricane Hilda.*

13. U.S. Army Corps of Engineers, *Report on Hurricane Betsy.*

14. "Nearly Half a Million People Beat Betsy to Safe Area"; "House Votes Flood Curbs"; U.S. Congress, House of Representatives, *Hurricane Betsy Disaster of September 1965.*

15. "President Promises Help."

16. "McKeithen Asks $10 Million Aid."

17. U.S. Congress, House of Representatives, *Hurricane Betsy Disaster.*

18. "Insurance Bill Speed Is Urged," and "Betsy Damage Hearings Due in N.O., Baton Rouge."

19. U.S. Army Corps of Engineers, *Disaster Operation*, quote at p. 2. Also U.S. Army Corps of Engineers, *Report on Hurricane Camille*, and Zebrowski and Howard, *Category 5*.

20. U.S. Congress, House of Representatives, *Hurricane Betsy Disaster*, 3–7.

21. *Ibid.* For the state as a whole, the governor estimated that damages could easily exceed $1 billion (32), and New Orleans Mayor Victor Schiro claimed that the city sustained property damage totaling half a billion dollars (79).

22. *Ibid.*, 7.

23. *Ibid.*, quote 10, and 10–12.

24. *Ibid.*, 32–33.

25. *Ibid.*, 79.

26. *Ibid.*, 133.

27. *Ibid.*, 84.

28. *Ibid.*, 82.

29. *Ibid.*, 85.

30. *Ibid.*, 35–39.

31. U.S. Congress, House of Representatives, *Southeast Hurricane Disaster (Hurricane Betsy)*. Insurance became a prominent issue in these hearings.

32. Platt, *Land Use and Society*, 425–27.

33. Colten, *Unnatural Metropolis*, 155–60. At the time Katrina struck in August 2005, Louisiana had significantly higher flood insurance subscribership rates than the national average. One report indicated that 64 percent of flooded homes had flood insurance. "Analysis: Insurance for Floods Prevalent."

34. Graham and Nunn, *Meteorological Considerations*, 1–2.

35. U.S. Army Corps of Engineers, *Interim Survey Report*, 24–30.

36. U.S. Army Corps of Engineers, *Final Environmental Impact Statement*, I-2. There were upward adjustments in levee heights from the interim report to the final design memoranda, but the wind speed for the standard project hurricane remained 100 mph. Corps engineers indicate that the standard underwent change. See also U.S. Army Corps of Engineers, *Performance Evaluation, The Hurricane Protection System*, III-11 to III-13. Solieau, personal interview, and Seale, personal interview.

37. U.S. Army Corps of Engineers, *Interim Survey Report*, 35–36.

38. Ibid., and U.S. Department of the Interior, *Detailed Report on Hurricane Study Area No. 1*.

39. U.S. Army Corps of Engineers, *Interim Survey Report*, 41–43.

40. *Ibid.*, 42.

41. *Ibid.*, 43.

42. *Ibid.*, 35, 38, and 41; Department of the Interior, *Detailed Report on Hurricane Study Area No. 1*; and U.S. Army Corps of Engineers, Waterways Experiment Station, *Effects on Lake Pontchartrain*.

43. U.S. Army Corps of Engineers, *Interim Survey Report*, 48–50.

44. *Ibid.*, 60–62.

45. Leander H. Perez (Plaquemines Parish Judge) to F. Edward Hebert (Louisiana Congressman), correspondence, September 18, 1963; F. Edward Hebert (Louisiana Congressman) to Judge Leander Perez (Plaquemines Parish Judge), September 24, 1963; and St. Bernard Parish Police Jury Resolution, November 5, 1963, all in U.S. Army Corps of Engineers, New Orleans District, National Archives, RG 77, Box 20, Survey Report Files, Fort Worth, Texas.

46. Secretary of the Army, *Letter*, 1965.

47. U.S. Congress, House of Representatives, *Flood Control Act of 1965*.

48. U.S. Army Corps of Engineers, *New Orleans Projects—Actual Costs*.

49. Kenneth J. Lesieur (Chairman, Citizens Committee for Hurricane Flood Control) to Col. T. J. Bowen (USACE, NOD), November 24, 1965, Interagency Performance Evaluation Task Force, https://ipet.wes.army.mil/, Office Files, LP&V HPP, Correspondence 1965–1977, viewed June 29, 2006.

50. U.S. Army Corps of Engineers, *Status Reports-Surveys*.

51. U.S. Army Corps of Engineers, *Public Hearing on St. Bernard Parish*.

52. Department of the Army, Office of the Chief of Engineers, *Annual Report 1968*, 427–28.

53. *Ibid.*, 428.

54. Department of the Army, *Annual Report 1970*, 403.

55. *Ibid.*, 401.

56. U.S. Department of the Interior, *Detailed Report*, 23; and U.S. Army Corps of Engineers, *Effects on Lake Pontchartrain*.

57. U.S. Army Corps of Engineers, *Final Environmental Impact Statement*, I-9 to I-11.

58. Landry, personal interview.

59. *Graci et al. v. U.S.*, 301 F. Supp., 947 (1969).

60. Soileau, personal interview.

61. *Graci et al. v. U.S.*, 435 F. Supp. 189 (1977).

62. Seale, personal interview, and Mazmanian and Neinaber, *Can Organizations Change?*, 79–90.

CHAPTER 4. LAKE PONTCHARTRAIN AND VICINITY HURRICANE PROTECTION, 1970–1990

1. Sands, personal interview, and Naomi, personal interview.

2. U.S. Army Corps of Engineers, *Design Memorandum 2, Lake Pontchartrain Barrier Plan*, I-6.

3. *Ibid.*, I-7, II–1. Also see U.S. Army Corps of Engineers, *Design Memorandum 1, Hydrology and Hydraulic Analysis, Part 1*.

4. U.S. Army Corps of Engineers, *Performance Evaluation*, v. IV: *The Storm*, IV-33.

5. U.S. Army Corps of Engineers, *Design Memorandum 2, Lake Pontchartrain Barrier* , III-6.

6. *Ibid.*, I-7.

7. *Ibid.*, III-2.

8. *Ibid.*, III-3.

9. Department of the Army, *Annual Report* 1970, 402.

10. Department of the Army, *Annual Report, 1974*, II-30.

11. U.S. Army Corps of Engineers, *Performance Evaluation, The Storm*, IV-39.

12. U.S. Army Corps of Engineers, *Design Memorandum 1, Hydrology and Hydraulic Analysis, Part 3, Lakeshore*, 1–2.

13. U.S. Army Corps of Engineers, *Design Memorandum 2, General Design, Citrus*, 10.

14. *Ibid.*, 74–75.

15. U.S. Army Corps of Engineers, *Design Memorandum 2, General Design, Supplement 5B.*

16. *Ibid.*, 7–8 and 21.

17. *Ibid.*, 11.

18. *Ibid.*, 16–18.

19. Department of the Army, *Annual Report 1975*, II-32.

20. U.S. Army Corps of Engineers, *Design Memorandum 1, Part 1, Hydrology and Hydraulic Analysis-Chalmette*, 28–29.

21. U.S. Army Corps of Engineers, New Orleans District, *Design Memorandum 3, General Design*, 15.

22. U.S. Army Corps of Engineers, *Design Memorandum 1, Part 1, Hydrology and Hydraulic Analysis-Chalmette*, 24.

23. Shortly after landfall, Hurricane Katrina winds across much of southeast Louisiana exceeded 98 mph—near or above the 100 mph design for the Chalmette-area levees. High-water marks topped 15 feet in this area. U.S. Army Corps of Engineers, *Performance Evaluation, The Storm*, IV-38 and IV-48. https://ipet.wes.army.mil/, viewed September 16, 2006, IV-39 and IV-49.

24. U.S. Army Corps of Engineers, *Design Memorandum 3, Lake Pontchartrain and Vicinity, Louisiana, Chalmette Area Plan, General Design*, 6.

25. *Ibid.*, 9–10. Responsibilities for land acquisition and rights for the ponding area resided with the cooperating local units, p. 36.

26. *Ibid.*, 63.

27. *Ibid.*, 8.

28. U.S. Army Corps of Engineers, *Design Memorandum 1, Part 1, Hydrology and Hydraulic Analysis-Chalmette*, 6.

29. U.S. Army Corps of Engineers, *Design Memorandum 3, Chalmette Extension*, 1.

30. Chief of Engineers to Division Engineer, Mississippi Valley Division, cover memo, October 21, 1968; U.S. Army Corps of Engineers, *Design Memorandum 3, Chalmette Extension.*

31. Department of the Army, *Annual Report 1975*, II-31 to II-32.

32. U.S. Army Corps of Engineers, *Lake Pontchartrain, Louisiana, and Vicinity, General Design Memorandum 2, Supplement 1, Rigolets Control Structure*, 27.

33. *Ibid.*, 11–13, 16–17, and 29.

34. *Ibid.*, 20.

35. *Ibid.*, 50.

36. U.S. Army Corps of Engineers, *Lake Pontchartrain, Louisiana, and Vicinity, General Design Memorandum 2, Supplement 3, Chef Menteur Pass Complex*, 10.

37. *Ibid.*, 12.

38. U.S. Fish and Wildlife Service to District Engineer, letter, May 15, 1968, in U.S. Army Corps of Engineers, *Lake Pontchartrain, Chef Menteur Pass Complex*, appendix B.

39. Orleans Levee District to Colonel Bowen, New Orleans District, correspondence, February 22, 1967, and Louisiana Department of Public Works to New Orleans District Engineer, correspondence, February 8, 1967, in U.S. Army Corps of Engineers, *Lake Pontchartrain, Chef Menteur Pass Complex*, appendix B.

40. U.S. Army Corps of Engineers, New Orleans District, *Lake Pontchartrain, Louisiana, and Vicinity, General Design Memorandum 2, Supplement 6, St. Charles Parish Lakefront Levee*, 6 and 24.

41. *Ibid.*, 6–7.

42. *Ibid.*, 34 and quote on 36.

43. U.S. Army Corps of Engineers, New Orleans District, *Review of Reports, St. Bernard Parish, Louisiana*, 17.

44. *Ibid.*, 38–40.

45. Levy to Bowen, correspondence; Bowen to Levy, correspondence. Litigation over floodings attributed to MRGO, see *Graci et al. v. U.S. of America*, et al., 435 F. Supp. 189 (1977).

46. Orleans Board to New Orleans District, correspondence, February 22, 1967; and Louisiana Department of Public Works to New Orleans District, correspondence, February 8, 1967, in *ibid.*

47. *Graci et al. v. U.S. of America*, 435 F. Supp. 189 (1977), Soileau, personal interview; and Bowen to Levy, correspondence.

48. Milton Dupuy, President, Orleans Levee Board, to Lt. Gen. William Cassidy, Chief of Engineers, telegram, December 4, 1968, NARA, RG 77, 72A, 3429, Box 1, Lake Pontchartrain folder, College Park, Maryland.

49. Hall to Dupy, correspondence.

50. Department of the Army, Corps of Engineers, *Annual Report, 1969*, 415–16.

51. Several letters forwarded from the local congressman, Hale Boggs, prompted a reply from the district engineer, Thomas Bowen, that he would review the recommendations. See Pilney to Boggs, correspondence; and Sivori to Boggs, correspondence. Reply from Bowen to Boggs, correspondence.

52. U.S. Army Corps of Engineers, *Design Memorandum 2, General Design, Supplement 5B*, 4–5.

53. U.S. Army, Corps of Engineers, *Record of Public Meeting*, 38.

54. Platt, *Land Use and Society*, 404–7. For a lengthy discussion of the Corps' incorporation of environmental concerns into its planning, see Mazanain and Nienaber, *Can Organizations Change?*

55. U.S. Army Corps of Engineers, *Lake Pontchartrain, Louisiana, and Vicinity Hurricane Protection Project*, i-ii.

56. U.S. Army Corps of Engineers, *Lake Pontchartrain, Louisiana, and Vicinity Hurricane Protection Project, Public Hearing, February 22, 1975*, 51.

57. Speaking for the Corps were Richard Richter, Stanley Shelton, and Glen Montz; *ibid.*, 17–42.

58. Harold Hart in *ibid.*, 53–59.

59. Arthur Theis in *ibid.*, 63–71, quotes on 67.

60. In the weeks after Katrina, ill-informed commentators made the case that environmentalists had blocked what would have been a more effective hurricane protection system. Environmentalists were allied in their efforts with pro-business groups, and much of the flooding in eastern New Orleans had nothing to do with the scrapped barrier plan. Edward Scoggin in *ibid.*, 75–100.

61. Frank Cusimano in *ibid.*, 101–7.

62. David Martin in *ibid.*, 194–98.

63. Joseph Burgess in *ibid.*, 110–19, quotes at 113 and 115.

64. David Levy in *ibid.*, 161–65.

65. Glenn Mercadal in *ibid.*, 168–71, quote at 171.

66. Cliff Danby in *ibid.*, 172–78.

67. William Fontenot in *ibid.*, 249–53, quote at 252. In fact, much of the New Orleans east section is now a wildlife refuge, see Colten, *Unnatural Metropolis*, esp. ch. 6.

68. Edgar Veillon in U.S. Army Corps of Engineers, *Public Hearing: The Lake Pontchartrain, Louisiana, and Vicinity Hurricane Protection Project, February 22, 1975*, 178–85, quote at 180.

69. Art Crowe in *ibid.*, 186–89.

70. Michael Tritico in *ibid.*, 238–43, quote at 240.

71. Earl Colomb in *ibid.*, 150–59.

72. William Gilmore in *ibid.*, 166–68.

73. Guy LeMieux in *ibid.*, 131–36.

74. Weston Strauch in *ibid.*, 190–94.

75. *Ibid.*, 1–12.

76. Heiberg, personal interview.

77. *Save our Wetlands v. Rush*, et al., 614 F. 2nd 1296 (1980).

78. U.S. Army Corps of Engineers, *Lake Pontchartrain, Louisiana, and Vicinity Hurricane Protection Project, Reeevaluation Study, Vol. 1, Main Report*, 1–2. Order, *Save our Wetlands et al. v. Early J. Rush II et al.*, Civil Action 75-3710, United States District Court, Eastern District of Louisiana, December 30, 1977, and Order, *Save our Wetlands et al. v. Early J. Rush II et al.*, Civil Action 75-3710, United States District Court, Eastern District of Louisiana, March 10, 1978.

79. U.S. Army Corps of Engineers, *Documentation of Public Meeting*, 7.

80. U.S. Army Corps of Engineers, *Lake Pontchartrain, Louisiana, and Vicinity Hurricane Protection Project, Reeevaluation Study, Vol. 1, Main Report*, 126.

81. *Ibid.*, 124.

82. Livingston to McGinnis, correspondence.

83. U.S. Army Corps of Engineers, *Lake Pontchartrain, Louisiana, and Vicinity Hurricane Protection Project, Reevaluation Study, Vol. 1, Main Report*, 132–34, quote on 133.

84. Comptroller General of the United States, *Report to Congress*, 15–16.

85. *Ibid.*, 21.

86. *Ibid.*, 21–22.

87. U.S. General Accounting Office, *Report to the Secretary of the Army*, 1.

88. *Ibid.*, 2–4.

89. *Ibid.*, 2–5.

90. Comptroller General of the United States, *Report to Congress*, 2 and 12. Also, U.S. Army Corps of Engineers, *1975 Annual Report*, II-30 to II-32.

91. U.S. General Accounting Office, *Report to the Secretary of the Army*, 8.

92. U.S. Army Corps of Engineers, *1983 Annual Report*, II-16 to II-17. Also, U.S. Army Corps of Engineers, *Status Report of New Orleans District Projects 1983*, 21.

93. U.S. Army Corps of Engineers, *Status Report of New Orleans District Projects 1990*, 6–7.

94. U.S. Department of the Army, *Annual Report 1989*, II-7.

95. Robert Guizerix, Minutes of Meeting: Lake Pontchartrain Outfall Canals Outfall Butterfly Valves, October 9, 1986, IPET Web Site, Lake Pontchartrain and Vicinity, Office Files, Miscellaneous, https://ipet.wes.army.mil/, viewed July 3, 2006; Leech, *Hurricane Protection Structure*; and Bottin and Mize, *Effects of Wave Action*.

96. U.S. Army Corps of Engineers, *Orleans Avenue Outfall Canal*, 53–54 and 58; and U.S. Army Corps of Engineers, *London Avenue Outfall Canal*, 61 and 66.

97. U.S. Army Corps of Engineers, *Orleans Parish-Jefferson Parish, 17th Street Outfall Canal*, 9–10, and 35.

98. Zebrowski and Howard, *Category 5*.

99. U.S. Army Corps of Engineers, *Report on Hurricane Camille*, 63–68.

100. *Ibid.*, 78–89.

101. *Ibid.*, 92–102.

102. *Ibid.*, 106–8.

103. U.S. Army Corps of Engineers, *History of Hurricane Occurences*, 36–37.

104. U.S. Army Corps of Engineers, *1985 Hurricanes Juan, Danny, Elena*, 66–72.

105. *Ibid.*, 1–54.

106. U.S. Army Corps of Engineers, *Budget Justifications*, Fiscal Year 1967.

107. Milton Dupuy, President Orleans Levee Board to Lt. Gen. William Cassidy, Chief of Engineers, telegram, December 4, 1968, NARA, RG 77, 72A, 3429, Box 1, Lake Pontchartrain folder, College Park, Maryland; and Hall to Dupy, correspondence.

108. District personnel consistently made this point. See Soileau, personal interview; Sands, personal interview; and Heiberg, personal interview.

109. Barry to Nixon, correspondence.

110. Comptroller General of the United States, *Report to Congress*, 10–12.

111. *Ibid.*, 13–14, quote 14.

112. *Ibid.*, 16.

113. U.S. General Accounting Office, *Report to the Secretary of the Army*, 2.

114. U.S. Army Corps of Engineers, *New Orleans Projects-Actual Costs.*

CHAPTER 5. PROTECTING THE DELTA, THE WEST BANK, AND THE COAST

1. Secretary of the Army, *Letter*, and U.S. Army Corps of Engineers, *New Orleans to Venice, Louisiana, Design Memorandum 1, General Design Reach 1*, 1.

2. U.S. Army Corps of Engineers, *New Orleans to Venice, Louisiana, Design Memorandum 1, General Design Reach 1*, A2.

3. *Ibid.*, 9.

4. *Ibid.*, B6–B9.

5. *Ibid.*, 14–15.

6. U.S. Army Corps of Engineers, *New Orleans to Venice, Louisiana Design Memorandum 1, Supplement 5 rev.*, 1–2 and 19–20.

7. *Ibid.*, 6–7 and 11–12.

8. *Ibid.*, 46.

9. *Ibid.*, 46 and 82.

10. U.S. Army Corps of Engineers, *New Orleans to Venice, Louisiana, Design Memorandum 1, General Design Reach 1*, B20–B21.

11. *Ibid.*, 5–10. Sand core and clay-cap levees were well known in engineering circles and served as the construction technique for many of the polders in the Netherlands since the early twentieth century.

12. U.S. Fish and Wildlife Service to U.S. Army Corps of Engineers, correspondence, November 29, 1969.

13. U.S. Army Corps of Engineers, *New Orleans to Venice, Louisiana, Design Memorandum 1, General Supplement 3, Reach C*, 1–7.

14. *Ibid.*, 17–19.

15. U.S. Army Corps of Engineers, *New Orleans to Venice, Louisiana, Design Memorandum 1, General Design Supplement 4, Reach B2*, 25–28.

16. *Ibid.*, 35–41.

17. U.S. Army Corps of Engineers, *New Orleans to Venice, Louisiana, Design Memorandum 1, General Supplement 3, Reach C*, A and 6–7.

18. *Ibid.*, 7–19.

19. *Ibid.*, quote at 24.

20. *Ibid.*, 35–38.

21. U.S. Army Corps of Engineers, *New Orleans to Venice, Louisiana, Hurricane Protection Project Final Supplemental Environmental Impact Statement, Supplement 2, Barrier Features*, 92.

22. U.S. Army Corps of Engineers, *New Orleans to Venice, Louisiana, Hurricane Protection: Final Environmental Impact Statement*, 11–15.

23. *Ibid.*, 15.

24. *Ibid.*, 16.

25. Department of the Army, *Annual Report 1973*, II-30.

26. Department of the Army, *Annual Report 1976*, II-33.

27. Department of the Army, *Annual Report 1982*, II-21.

28. Department of the Army, *Annual Report 1985*, II-12.

29. Department of the Army, *Annual Report 1989*, II-9.

30. Department of the Army, *Annual Report 1992*, II-7.

31. Department of the Army, *Annual Report 1997*, II-7.

32. Barnes, "Four Miles of Levee."

33. Bazile, "Corps Tells Parish to Stop Work on Levee."

34. Department of the Army, *Annual Report 2004*, II-7, and U.S. Army Corps of Engineers, *New Orleans to Venice Fact Sheet.*

35. U.S. Army Corps of Engineers, *New Orleans Projects-Actual Costs Spreadsheet.*

36. "Corps Sees Its Resources Siphoned Off," and "Critics Say Change Priorities."

37. U.S. Army Corps of Engineers, *New Orleans to Venice Fact Sheet.*

38. U.S. Army Corps of Engineers, *General Design Memorandum 1 (Reduced Scope)*, A-1.

39. Colten, *Unnatural Metropolis*, 155–61, and URS Company, *West Bank Master Drainage Study.*

40. U.S. Army Corps of Engineers, *Report on Hurricane Betsy*, 35–37; and U.S. Army Corps of Engineers, *Hurricane Betsy: After Action Report.*

41. U.S. Army Corps of Engineers, *West Bank of the Mississippi River in the Vicinity of New Orleans*, 6–7; and National Park Service, *Jean Lafitte National Historical Park.*

42. U.S. Army Corps of Engineers, *West Bank of the Mississippi River in the Vicinity of New Orleans*, 7.

43. *Ibid.*, 8; and "Disputes Delay Hurricane Levees."

44. "Jeff Trying to Hurry New Levees."

45. "Council Proposes Jeff Build Levee," and "Protecting West Bank."

46. Vincent, *Letter to the Editor.*

47. "Jeff Council Amends Hurricane Line"; "'No Growth' Line in Jeff Unresolved"; and "West Bank Hurricane Levee Plan Criticized."

48. "West Bank Levee Battle Restaged at Corps Hearing."

49. "Jeff Levee Plan Turned Down, Another Urged."

50. "Hurricane Could Overrun Jeff Levees."

51. U.S. Army Corps of Engineers, *West Bank of the Mississippi River in the Vicinity of New Orleans*, 8–9; Burdine, personal interview; and Sands, personal interview.

52. U.S. Army Corps of Engineers, *1985 Hurricanes Juan, Danny, Elena*, 42.

53. *Ibid.*, 44–47.

54. *Ibid.*, 48.

55. U.S. Army Corps of Engineers, *West Bank of the Mississippi River*, EIS sec. 1–2.

56. U.S. Army Corps of Engineers, *West Bank of the Mississippi River*, EIS sec. 6–7.

57. U.S. Army Corps of Engineers, *General Design Memorandum 1 (Reduced Scope)*, xi.

58. *Ibid.*, 7.

59. *Ibid.*, 10–11.

60. *Ibid.*, 56–57.

61. U.S. Army Corps of Engineers, *Design Memorandum 2, East and West of Algiers Canal*, 1–4.

62. "Parish Builds Levee in Marsh"; "Hurricane Levee Plans Reviewed Today"; "Levee Land Battles Brew."

63. "Harvey Canal Floodgate Studied"; and "Harvey Canal Flood Plan Urged."

64. *Memo for the Record*, and U.S. Corps of Engineers, *Fact Sheet*.

65. Yenni to Diffley, correspondence.

66. Lee, "Corps Told to Hurry Levee Work."

67. U.S. Army Corps of Engineers, *Feasibility Review Conference*; Mississippi Valley Division Commander to Diffley, Memorandum; and quote in Edwards, and additional signees, to Diffley, correspondence.

68. "West Bank Levee Starts Amid Fanfare."

69. U.S. War Department, *Annual Report 1992*, II-9.

70. Boffone to Diffley, correspondence.

71. Schroeder to Commander Lower Mississippi Valley Division, memo.

72. U.S. War Department, *Annual Report 1995*, II-7.

73. Louwaigie, "Eye of the Storm."

74. U.S. War Department, *Annual Report 2000*, II-9.

75. Barbier, "Harvey Canal Drainage Contract Awarded."

76. U.S. War Department, *Annual Report 2004*, II-8; and U.S. Army Corps of Engineers, "West Bank and Vicinity, New Orleans, Louisiana, Hurricane Protection Project, Project Fact Sheet.

77. "Levee Land Battles Brew," and "Council Agrees to Pay for New Wetlands."

78. U.S. Army Corps of Engineers, *New Orleans Projects-Actual Costs Spreadsheet*.

79. Spohrer to Broussard, correspondence and attached Budgetary Procedures.

80. Lee, "Congressmen Will Be Pressed to Push for Hurricane Levee."

81. Louwaigie, "Eye of the Storm."

82. Young, "Hurricane Levee Should Be Done by 2004."

83. Barbier, "Levee Strapped for Cash."

84. Cross, "Mean Season."

85. "Critic Says Change Priorities."

86. Meyer-Arendt, "Grand Isle, Louisiana, Resort Cycle"; and U.S. Army Corps of Engineers, *Grand Isle and Vicinity Louisiana: Final Environmental Impacts Statement*, II-1.

87. U.S. Army Corps of Engineers, *Grand Isle and Vicinity Louisiana: Final Environmental Impacts Statement*, II-4; and U.S. Army Corps of Engineers, *Report on Hurricane Betsy*, 35–37.

88. U.S. Army Corps of Engineers, *Grand Isle and Vicinity Louisiana: Final Environmental Impacts Statement*, II-4.

89. U.S. Army Corps of Engineers, *Grand Isle and Vicinity Louisiana: Phase 1, General Design Memorandum*, 26–27.

90. U.S. Army Corps of Engineers, *Grand Isle and Vicinity Louisiana: Final Environmental Impacts Statement*, III-1.

91. *Ibid.*, III-2 and frontispiece.

92. U.S. Army Corps of Engineers, *Grand Isle and Vicinity Louisiana: Phase 1, General Design Memorandum*, 49.

93. *Ibid.*, 31.

94. *Ibid.*, 71–78 and 92.

95. "Grand Isle Levee Plans Underway."

96. U.S. Army Corps of Engineers, *Grand Isle and Vicinity Louisiana: Phase 1, General Design Memorandum*, 84 and 98.

97. U.S. Department of the Army, *Annual Report 1985*, II-9.

98. Torres and Nelson, "Unhappy Returns."

99. U.S. Army Corps of Engineers, *Grand Isle and Vicinity, Louisiana: Project Fact Sheet.*

100. U.S. Army Corps of Engineers, *Grand Isle, Louisiana and Vicinity (Larose to Vicinity of Golden Meadow), Design Memorandum 1, General Design*, 10–11.

101. *Ibid.*, 6–7.

102. *Ibid.*, 4–5, 47, and Appendix A.

103. *Ibid.*, 10.

104. U.S. Department of the Army, *Annual Report 1982*, II-20.

105. U.S. Department of the Army, *Annual Report 1985*, II-10 to II-11, and *Annual Report 1989*, II-7 to II-8.

106. U.S. Department of the Army, *Annual Report 1994*, II-5.

107. U.S. Department of the Army, *Annual Report 2000*, II-7, and *Annual Report 2004*, II-6.

108. U.S. Army Corps of Engineers, "Larose to Golden Meadow, Louisiana Hurricane Protection Project: Project Fact Sheet."

CHAPTER 6. DEVELOPMENTS TO THE EVE OF KATRINA, 1990–2005

1. U.S. Department of Commerce, *Hurricane Andrew*, 54–55.

2. *Ibid.*, 128.

3. National Oceanic and Atmospheric Administration, "Preliminary Report: Hurricane Andrew."

4. U.S. Department of Commerce, *Hurricane Andrew*, E1–E7.

5. "Andrew Weakens after Slamming Louisiana," *Washington Post*, 27 August 1992, A1.

6. "Hurricane Readiness Upgraded Since Andrew."

7. "Seawall Project off the Ground."

8. Secretary of the Army, *Annual Report 1994*, II-4.

9. National Oceanic and Atmospheric Administration, "Preliminary Report, Hurricane Danny."

10. "Grand Isle Hit from 'Back'"; "Repairs Set for Buras Harbor Hit by Danny"; and "Danny's Damage in the Millions."

11. National Oceanic and Atmospheric Administration, "Preliminary Report, Hurricane Georges."

12. "Tallying the Damage"; "Georges Takes Toll on Lakefront."

13. "Hurricane Silts New Orleans Port."

14. Hecht, "Geography Shifts in the Wake of the Storm."

15. Helman, "Port in a Storm."

16. "Isidore Drenching New Orleans Area"; "Adios Isidore"; "Money Down the Drain."

17. National Oceanic and Atmospheric Administration, "Tropical Cyclone Report Hurricane Isidore."

18. Colten, *Unnatural Metropolis*, 153.

19. "Levee Strapped for Cash."

20. National Oceanic and Atmospheric Administration, "Tropical Cyclone Report: Hurricane Lili"; "Lili Proves More Wild, But Not as Wet"; and Torres and Nelson, "Unhappy Returns."

21. "Levee Strapped for Cash."

22. National Oceanic and Atmospheric Administration, "Tropical Cyclone Report: Tropical Storm Bill."

23. National Oceanic and Atmospheric Administration, "Tropical Cyclone Report: Hurricane Ivan."

24. National Oceanic and Atmospheric Administration, "Tropical Cyclone Report: Tropical Storm Matthew."

25. Keim and Muller, "Temporal Fluctuations of Heavy Rainfall Magnitudes" and "Frequency of Heavy Rainfall Events."

26. Colten, *Unnatural Metropolis*, 149–57.

27. *Ibid.*, 151–59.

28. U.S. Army Corps of Engineers, *Jefferson and Orleans Parishes*, 23–35.

29. U.S. Army Corps of Engineers, "Post Flood Report," 3.

30. U.S. Army Corps of Engineers, *Southeast Louisiana Project, Jefferson, Orleans, and St. Tammany Parishes, Louisiana: Technical Report*, 1–6.

31. *Ibid.*, 21.

32. U.S. Army Corps of Engineers, *Southeast Louisiana Project: Jefferson Parish Technical Report*, and *Southeast Louisiana Project: Orleans Parish Technical Report.*

33. Secretary of the Army, *Annual Report 1997*, II-7.

34. "Slidell Drainage, Flood-Control Work Hit Snags."

35. "Flood Projects Fall Short of Money."

36. *Ibid.*

37. "Drainage Projects Will Be Suspended."

38. "Pumping Station to Be Completed Soon."

39. "Flood Projects May Stay Afloat."

40. Secretary of the Army, *Annual Report 1997*, II-7.

41. Colten, *Unnatural Metropolis*, 153–54.

42. U.S. Army Corps of Engineers, *Lake Pontchartrain and Vicinity, Lake Pontchartrain High Level Plan, Jefferson Parish Lakefront Levee, Design Memorandum 17*, 10.

43. *Ibid.*, 83–84.

44. U.S. Army Corps of Engineers, *Lake Pontchartrain and Vicinity, Lake Pontchartrain High Level Plan, St. Charles Parish North of Airline Highway, Design Memorandum 18*, 9, 15, 37.

45. U.S. Army Corps of Engineers, "Key Feature of St. Charles Hurricane Protection Ready for Duty."

46. "Levee Money Falling Short."

47. U.S. Army Corps of Engineers, *Lake Pontchartrain, Louisiana and Vicinity, High Level Plan, Design Memorandum 22: Orleans Parish Lakefront Remaining Work*, xv and 2.

48. Secretary of the Army, *Annual Report 1997*, II-6.

49. The drainage system in Orleans Parish differed substantially from the suburban system in Jefferson Parish. The Orleans Parish system represented an older design that placed the pumps at the low points in the city in order to lift water over the Metairie and Gentilly ridges into canals that allowed it to drain into Lake Pontchartrain. Jefferson Parish placed its pumps at the lakefront behind the massive earthen levees and avoided the need for parallel levees along the canals.

50. *Orleans Avenue Outfall Canal, Memorandum 19, General Design*, 61; *London Avenue Canal, Orleans Parish: Design Memorandum 19A, General Design*, v. I, cover memo, 3, 61, and 68; U.S. Army Corps of Engineers, *Orleans Parish-Jefferson Parish, 17th St. Outfall Canal, General Design, Design Memorandum 20*, executive summary.

51. U.S. Army Corps of Engineers, *Orleans Parish-Jefferson Parish, 17th St. Outfall Canal, General Design, Design Memorandum 20*, 8–10. As of 1989, the Corps still recommended installing butterfly gates on the London Avenue Canal. *London Avenue Outfall Canal, Orleans Parish: Design Memorandum 19A, General Design*, v. I, 7–9. *Orleans Avenue Outfall Canal, Design Memorandum 19, General Design*, 8.

52. "$6.7 Million Levee Bill Too Much for St. Bernard"; "Victory Declared on Levee Debt"; and "Parish's Levee Debt Resurfaces."

53. "Work Begins to Raise Lake Pontchartrain Levee."

54. "Airport's Sinking Levee to Be Restored."

55. "Feds Seek Input on Levee Project."

56. "Critic Says Change Priorities."

57. "Flood Control Program Will Go On" and "Cash Is Tight for Storm Protection."

58. Burdine, interview.

59. Naomi, interview.

60. "Army Corps Planning Category 5 Hurricane Protection Study."

61. "Hurricane Study Funds Hard to Find."

62. U.S. Army Corps of Engineers, "Memorandum to Commander, Mississippi Valley Division," 1–2.

63. *Ibid.*, 3–4.

64. *Ibid.*, 6–7.

65. *Ibid.*, 7. Post-Katrina measurements determined some levee sections were not up to design height even though officials thought they were.

66. *Ibid.*, 8–9.

67. *Ibid.*, 9.

68. *Ibid.*, 10–11.

69. *Ibid.*, 7–8.

70. "Betsy a Big One But Wound Not Deep"; "Betsy Winds Topped 15' High"; "McKeithen Asks $10 Million Aid."

71. U.S. Army Corps of Engineers, Federal Emergency Management Agency, and the National Weather Service, *Southeast Louisiana Hurricane Preparedness Study*, III-3.

72. *Ibid.*, IV-5 to IV-6.

73. "Left Behind."

74. U.S. Army Corps of Engineers, *Un-watering Plan.*

75. U.S. Army Corps of Engineers, *Abbreviated Transportation Model*, 2.

76. Fischetti, "Drowning New Orleans"; "Nothing's Easy For New Orleans Flood Control"; Cohen, "If the Big One Hits, New Orleans Could Disappear." FEMA held its national flood conference in New Orleans in May 2002.

77. McQuaid and Schleifstein, "Washing Away".

78. Burdine, interview; Naomi, interview.

79. Louisiana Water Resources Research Institute, "Louisiana Coastal Images Archive."

80. Laska, "What If Hurricane Ivan Had Not Missed New Orleans," and Williams, Penland, and Sallenger, *Atlas of Shoreline Changes.*

81. Louisiana Coastal Wetlands Conservation and Restoration Task Force, *Coast 2050*, 13.

82. "Corps Study to Evaluate Coastline Restoration Efforts."

83. U.S. Army Corps of Engineers, "Caernarvon Freshwater Diversion Project," and U.S. Army Corps of Engineers, "New Orleans Water Flows through Davis Pond."

84. U.S. Army Corps of Engineers, "West Bay Sediment Diversion."

85. Louisiana Coastal Wetlands Conservation and Restoration Task Force, *Coastal Wetlands Planning, Protection, and Restoration Acts*, 13.

CHAPTER 7. CONCLUSIONS

1. Several examples include Brinkley, *Great Deluge*; Van Heerden and Bryan, *Storm*; and Center for Public Integrity, *City Adrift*.

2. U.S. Army Corps of Engineers, *Performance Evaluation of the New Orleans and Southeast Louisiana Hurricane Protection System, Final Draft*; and Seed et al., "Investigation of the Performance of the New Orleans Hurricane Protection Systems"; and Team Louisiana, *Failure of the New Orleans Levee System.*

3. U.S. Congress, Select Bipartisan Committee, *Failure of Initiative*; U.S. Congress, Senate, Committee on Homeland Security and Governmental Affairs, *Hurricane Katrina*;

and U.S. Department of Homeland Security, Office of Inspector General, *Performance Review*; and U.S. Executive Office of the President, *Federal Response to Hurricane Katrina.*

4. U.S. Congress, Senate, *Lake Pontchartrain: Letter from the Secretary of the Army*, 1–2.

5. U.S. Army Corps of Engineers, *Interim Survey Report*, and Secretary of the Army, *Letter from the Secretary of the Army.*

6. "New Orleans Sitting Duck for Major Hurricane."

7. See Fischetti, "Drowning New Orleans"; and McQuaid and Schleifstein, "Washing Away."

8. "Army Corps Planning Category 5 Hurricane Protection Study."

9. U.S. Army Corps of Engineers, *Louisiana Coastal Protection and Restoration Technical Report.*

10. Shallat, "In the Wake of Hurricane Betsy."

11. Kelman, *River and Its City*, 19–49; and Colten, *Unnatural Metropolis*, 21–22.

12. Mazmanian and Neinaber, *Can Organizations Change*, 79–97, and Colten, *Unnatural Metropolis*, 109–25.

13. U.S. Army Corps of Engineers, "Levees and Floodwalls."

14. *Idem*, "Hurricane Protection Improvements."

15. *Idem*, "Existing Pumping Station Stormproofing."

16. *Idem*, "Planned Permanent Pump Station Locations."

17. *Idem*, "MRGO Report Sent to Congress."

18. Carter, *New Orleans Levees and Floodwalls*, 5–6.

19. U.S. Army Corps of Engineers, *Louisiana Coastal Protection and Restoration Technical Report*, 2.

20. *Idem*, *Performance Evaluation*, v. III, and "Risk and Reliability Report."

21. U.S. Army Corps of Engineers,

22. Order and Reasons, Katrina Canal Breaches Consolidated Litigation, CA No. 05-4182, 44–45, 533 F. Supp. 615 (January 2008).

23. *Tommaseo et al. v. U.S.*, and "Judge: Corps Can Be Sued for Flood."

BIBLIOGRAPHY

"An Act Relative to Roads, Levees, and the Police of Cattle." *Acts Passed Before the Second Session of the First Legislature of the Territory of Orleans.* New Orleans: Bradford and Anderson, 1807.

"Adios Isidore." *New Orleans Times Picayune*, September 27, 2002.

"Airport's Sinking Levee to Be Restored." *New Orleans Times Picayune*, June 22, 2003.

"Analysis: Insurance for Floods Prevalent," *Baton Rouge Advocate*, March 20, 2006.

"Andrew Weakens After Slamming Louisiana." *Washington Post*, August 27, 1992.

"Army Corps Planning Category 5 Hurricane Protection Study." *New Orleans Times Picayune*, November 4, 2002.

Arnold, Joseph L. *The Evolution of the 1936 Flood Control Act.* Fort Belvoir, Va.: U.S. Army Corps of Engineers, History Office, 1988.

"Backwater: Rear of City Inundated." *New Orleans Daily Picayune*, April 22, 1890.

Barbier, Sandra. "Harvey Canal Drainage Contract Awarded." *New Orleans Times Picayune*, October 25, 2002.

———. "Levee Strapped for Cash." *New Orleans Times Picayune*, October 19, 2002.

Barnes, Sam. "Four Miles of Levee Raised for Hurricane Protection." *Louisiana Contractor* 48, no. 9 (August 1999): 73.

Barry, Denis A. (Regional Planning Commission) to Richard M. Nixon (President of the United States). Correspondence, June 24, 1971, Lake Pontchartrain and Vicinity, Office Files, 1965–77, https://ipet.wes.army.mil/, viewed July 6, 2006.

Barry, John M. *Rising Tide: The Great Mississippi Flood of 1927 and How It Changed America.* New York: Simon & Schuster, 1997.

Bazile, Karen. "Corps Tells Parish to Stop Work on Levee." *New Orleans Times Picayune*, March 20, 2003.

"Betsy a Big One But Wound Not Deep." *New Orleans Times Picayune*, September 11, 1965.

"Betsy Damage Hearings Due in N.O., Baton Rouge." *New Orleans Times Picayune*, September 21, 1965.

"Betsy Winds Topped 15' High." *New Orleans Times Picayune*, September 19, 1965.

Bixel, Patricia, and Elizabeth Turner. *Galveston and the 1900 Storm*. Austin: University of Texas Press, 2000.

Boffone, Francis, (President of West Jefferson Levee District) to Col. Michael Diffley (District Engineer, New Orleans District Corps of Engineers). Correspondence, February 16, 1993. Project Management Files, folder 2, New Orleans District.

Bottin, Robert, and Marvin G. Mize. *Effects of Wave Action on a Hurricane Protection Structure for London Avenue Outfall Canal, Lake Pontchartrain, New Orleans, Louisiana, U.S. Army Corps of Engineers, Waterways Experiment Station, Miscellaneous Paper CERC-87-14*. New Orleans: U.S. Army Corps of Engineers, New Orleans District, 1987.

Bowen, Col. Thomas, (District Engineer, New Orleans District, Corps of Engineers) to Neville Levy. Correspondence, July 21, 1967. National Archives, RG 77, 71A-1971, Box 1, Lake Pontchartrain file, College Park, Maryland.

Bowen, Col. Thomas, (District Engineer, New Orleans District, Corps of Engineers) to Hale Boggs (Louisiana Congressman). Correspondence, April 24, 1968. National Archives, RG 77, 3429, Box 1, Lake Pontchartrain file, College Park, Maryland.

Brinkley, Douglas. *The Great Deluge: Hurricane Katrina, New Orleans and the Mississippi Gulf Coast*. New York: William Morrow, 2006.

Burby, Raymond J. "Hurricane Katrina and the Paradoxes of Government Disaster Policy." *Annals American Association for Political and Social Sciences* 604 (2006): 171–91.

Burdine, Carol. (U.S. Army Corps of Engineers, New Orleans District, Project Manager, West Bank Hurricane Protection Project.) Personal interview, January 31, 2006.

Burk and Associates. *East Bank Master Drainage Plan: Jefferson Parish Louisiana*, v. I. Gretna, La.: Jefferson Parish, 1980.

Camillo, Charles A., and Matthew T. Pearcy. *Upon Their Shoulders: A History of the Mississippi River Commission from Its Inception through the Advent of the Modern Mississippi River and Tributaries Project*. Vicksburg, Miss.: Mississippi River Commission, 2004.

Carter, Nicloe T. *New Orleans Levees and Floodwalls: Hurricane Damage Protection*. Washington, D.C.: Congressional Research Service.

"Cash Is Tight for Storm Protection." *New Orleans Times Picayune*, December 4, 2004.

Castonguay, Stephane. "The Production of Flood as Natural Catastrophe: Extreme Events and the Construction of Vulnerability in the Drainage Basin of the St. Francis River (Quebec), Mid-Nineteenth Century to Mid-Twentieth Century." *Environmental History* 12 (2007): 820–44.

Center for Public Integrity. *City Adrift: New Orleans Before and After Katrina*. Baton Rouge: Louisiana State University Press, 2007.

Cohen, Adam. "If the Big One Hits, New Orleans Could Disappear." *New York Times*, August 11, 2002.

Colten, Craig E. "Bayou St. John: Strategic Waterway of the Louisiana Purchase." *Historical Geography* 31 (2003): 21–30.

———, ed. *Transforming New Orleans and Its Environs: Centuries of Change.* Pittsburgh: University of Pittsburgh Press, 2000.

———. *An Unnatural Metropolis: Wresting New Orleans from Nature.* Baton Rouge: Louisiana State University Press, 2005.

Comptroller General of the U.S. *Report to Congress: Cost, Schedules, and Performance Problems of the Lake Pontchartrain and Vicinity, Louisiana Hurricane Protection Project.* Washington, D.C.: U.S. General Accounting Office, 1976.

"Corps Sees Its Resources Siphoned Off." *New Orleans Times Picayune*, April 24, 2004.

"Corps Study to Evaluate Coastline Restoration Efforts." *New Orleans Times Picayune*, January 3, 1988.

"Council Agrees to Pay for New Wetlands." *New Orleans Times Picayune*, August 6, 2002.

"Council Proposes Jeff Build Levee." *New Orleans Times Picayune*, September 4, 1980.

Cowdrey, Albert E. *Land's End: A History of the New Orleans District of the U.S. Army Corps of Engineers.* New Orleans: U.S. Army Corps of Engineers, New Orleans District, 1977.

"Critic Says Change Priorities." *Baton Rouge Advocate*, September 17, 2005.

Cross, Kim. "The Mean Season." *New Orleans Times Picayune*, May 31, 2003.

Cutter, Susan. *American Hazardscapes: The Regionalization of Hazards and Disasters.* Washington, D.C.: Joseph Henry Press, 2001.

———, et al. "Revealing the Vulnerability of People and Places: A Case Study of Georgetown County, South Carolina." *Annals of the Association of American Geographers* 90 (2000): 713–37.

"Danny's Damage in the Millions." *New Orleans Times Picayune*, July 20, 1997.

Davis, Donald W. "Historical Perspective on Crevasses, Levees, and the Mississippi River." In *Transforming New Orleans and Its Environs: Centuries of Change*, edited by Craig E. Colten, 84–106. Pittsburgh: University of Pittsburgh Press, 2000.

Davis, Mike. *Ecology of Fear: Los Angeles and the Imagination of Disaster.* New York: Henry Holt, 1998.

"Disputes Delay Hurricane Levees." *New Orleans Times Picayune*, January 9, 1979.

"Drainage Projects Will Be Suspended." *New Orleans Times Picayune*, February 19, 2004.

Dunn, Gordon E., and Banner I. Miller. *Atlantic Hurricanes.* Baton Rouge: Louisiana State University Press, 1960.

Dupuy, Milton, (President, Orleans Levee Board) to Lt. Gen. William Cassidy (Chief of Engineers). Telegram, December 4, 1968. National Archives, RG 77, 72A, 3429, Box1, Lake Pontchartrain folder, College Park, Maryland.

Edwards, Governor Edwin, et al. to Col. Mike Diffley (District Engineer). Correspondence, May 5, 1993. Project Management Files, Westwego to Harvey, folder 1, New Orleans District.

Elliott, D. O. *The Improvement of the Lower Mississippi River for Flood Control and Navigation.* Vicksburg, Miss.: U.S. Army Corps of Engineers, U.S. Waterways Experiment Station, 1932.

Elsner, James B., and A. Birol Kara. *Hurricanes of the North Atlantic: Climate and Society.* New York: Oxford, 1999.

"Evacuees Returning Home as Flood Problem Created by Flossy Improves." *New Orleans Times Picayune*, October 25, 1956.

"Feds Seek Input on Levee Project." *New Orleans Times Picayune*, February 1, 2003.

Fischetti, Mark. "Drowning New Orleans." *Scientific American* 285, no. 4 (2001): 77–85.

"Flood Control Program Will Go On." *New Orleans Times Picayune*, June 19, 2003.

"Flood Projects Fall Short of Money." *New Orleans Times Picayune*, June 5, 2002.

"Flood Projects May Stay Afloat." *New Orleans Times Picayune*, December 1, 2004.

"Georges Takes Toll on Lakefront." *New Orleans Times Picayune*, September 30, 2006.

Graci et al. v. U.S., 301 F. Supp. 947 (1969).

Graci et al. v. U.S., 435 F. Supp. 189 (1977).

Graham, Howard, and Dwight Nunn. *Meteorological Considerations Pertinent to Standard Project Hurricane, Atlantic and Gulf Coasts of the United States, National Hurricane Research Project Report No. 33.* Washington, D.C.: U.S. Department of Commerce, Weather Bureau, 1959.

"Grand Isle Hit from 'Back.'" *New Orleans Times Picayune*, August 5, 1997.

"Grande Isle Levee Plan Underway." *New Orleans Times Picayune*, August 25, 1979.

Gumprecht, Blake. *The Los Angeles River: Its Life, Death, and Possible Rebirth.* Baltimore: Johns Hopkins University Press, 1999.

Hall, Daniel, (USACE, Mississippi Valley Division) to Milton Dupuy (Orleans Levee Board). Correspondence, December 13, 1968. National Archives, RG 77, 72A, 3429, Box 1, Lake Pontchartrain folder, College Park, Maryland.

Hardee, T. S. 1878. *Topographic and Drainage Map of New Orleans*, facsimile edition. 1878; New Orleans: Historic New Orleans Collection, 2000.

Harrison, Robert W. *Alluvial Empire*, v. I: *A Study of State and Local Efforts toward Land Development in the Alluvial Valley of the Lower Mississippi River, Including Flood Control, Land Drainage, Land Clearing, Land Forming.* Little Rock: U.S. Department of Agriculture, Economic Research Service, 1961.

———. *Swamp Land Reclamation in Louisiana, 1849–1979: A Study of Flood Control and Land Drainage.* Baton Rouge, La.: U.S. Bureau of Agricultural Economics, 1951.

"Harvey Canal Floodgate Studied." *New Orleans Times Picayune*, September 8, 1989.

"Harvey Canal Flood Plan Urged." *New Orleans Times Picayune*, November 17, 1989.

Hecht, Jeff. "Geography Shifts in the Wake of the Storm." *New Scientist*, October 10, 1998, 16.

Heiberg, Col. E. R. III. Personal interview, February 16, 2006.

Helman, Christopher. "Port in a Storm." *Forbes*, October 3, 2005, 56.

"House Votes Flood Curbs." *New Orleans Times Picayune*, September 23, 1965.

"Hurricane Could Overrun Jeff Levees." *New Orleans Times Picayune*, June 23, 1984.

"Hurricane Levee Plans Reviewed Today." *New Orleans Times Picayune*, May 27, 1999.

"Hurricane Readiness Upgraded Since Andrew." *Baton Rouge Advocate*, September 2, 1994.

"Hurricane Silts New Orleans Port." *Engineering News-Record*, November 2, 1998, 20.

"Hurricane Study Funds Hard to Find," *New Orleans Times Picayune*, July 22, 2003.

"Insurance Bill Speed Is Urged." *New Orleans Times Picayune*, September 10, 1965.

"The Inundation." *New Orleans Daily Picayune*, April 23, 1890.

"Isidore Drenching New Orleans Area." *New Orleans Times Picayune*, September 26, 2002.

"Jeff Council Amends Hurricane Line." *New Orleans Times Picayune*, February 22, 1979.

"Jeff Hurricane Levee Plan Turned Down, Another Urged." *New Orleans Times Picayune*, June 21, 1984.

"Jeff Trying to Hurry New Levees." *New Orleans Times Picayune*, January 11, 1979.

"Judge: Corps Can Be Sued for Flood." May 2, 2008. http://www.wwltv.com/local/stories/wwl050208tpmrgo.cof81f80.html#.

Keim, Barry D., and Robert A. Muller. "Frequency of Heavy Rainfall Events in New Orleans, Louisiana, 1900 to 1991." *Southeastern Geographer* 33, no. 2 (1993): 159–71.

———. "Temporal Fluctuations of Heavy Rainfall Magnitudes in New Orleans Louisiana, 1871–1991." *Water Resources Bulletin* 28, no. 4 (1992): 721–30.

Kelman, Ari. *A River and Its City: The Nature of Landscape in New Orleans*. Berkeley: University of California Press, 2003.

Kidder, Tristam. "Making the City Inevitable: Native Americans and the Geography of New Orleans." In *Transforming New Orleans and Its Environs: Centuries of Change*, edited by Craig E. Colten, 9–21. Pittsburgh: University of Pittsburgh Press, 2000.

Landry, Victor A. (Corps of Engineers, retired). Personal interview, December 19, 2005.

Laska, Shirley. "What If Hurricane Ivan Had Not Missed New Orleans." *Natural Hazards Observer* 29 (2004): 1.

Lee, Vincent. "Congressmen Will Be Pressed to Push for Hurricane Levee." *New Orleans Times Picayune*, February 23, 1996.

———. "Corps Told to Hurry Levee Work." *New Orleans Times Picayune*, July 22, 1994.

Leech, J. *Hurricane Protection Structure for London Avenue Outfall Canal, Lake Pontchartrain, New Orleans, Louisiana: Hydraulic Model Investigation*. Vicksburg, Miss.: U.S. Army Engineer Waterways Experiment Station, 1987.

"Left Behind." *New Orleans Times Picayune*, June 24, 2002.

"Levee Land Battles Brew." *New Orleans Times Picayune*, March 2, 2000.

"Levee Money Falling Short." *New Orleans Times Picayune*, April 13, 2004.

"Levee Strapped for Cash." *New Orleans Times Picayune*, October 19, 2002.

Levy, Neville, to Colonel Thomas Bowen (District Engineer, New Orleans District). Correspondence, July 14, 1967. National Archives, RG77, 71A-1971, Box 1, Lake Pontchartrain file, College Park, Maryland.

Lewis, Peirce F. *New Orleans: The Making of an Urban Landscape*. Cambridge, Mass.: Ballinger, 1976.

"Lili Proves More Wild, But Not as Wet." *New Orleans Times Picayune*, October 4, 2002.

Livingston, Congressman Robert L., to Major General Charles I. McGinnis (U.S. Army Corps of Engineers, Director of Civil Works). Correspondence, February 15, 1978, IPET web site, MVN Records, https://ipet.wes.army.mil/, viewed July 6, 2006.

Louisiana Coastal Wetlands Conservation and Restoration Task Force. *Coast 2050: Toward a Sustainable Coastal Louisiana*. Baton Rouge: Louisiana Department of Natural Resources, 1998.

——. *Coastal Wetlands Planning, Protection, and Restoration Acts: A Response to Louisiana's Land Loss.* Baton Rouge: Louisiana Coastal Wetlands Conservation and Restoration Task Force, 2006.

Louisiana Department of Public Works to New Orleans District Engineer. Correspondence, February 8, 1967. In *U.S. Army Corps of Engineers, New Orleans District, Lake Pontchartrain, Louisiana, and Vicinity, General Design Memorandum 2, Supplement 3, Chef Menteur Pass Complex*, Appendix B. New Orleans: U.S. Army Corps of Engineers, New Orleans District, 1969.

Louisiana Water Resources Research Institute. "Louisiana Coastal Images Archive." http://www.lwrri.lsu.edu/1998_2002WEB/cia/html/cia.html, viewed July 1, 2006.

Louwaigie, Pam. "Eye of the Storm: Levee-Building Delays Leave West Bank Vulnerable." *New Orleans Times Picayune*, June 1, 1999.

"Lower Parishes Pushing Relief." *New Orleans Times Picayune*, October 26, 1956.

Mazmanian, Daniel, and Jeanne Nienaber. *Can Organizations Change: Environmental Protection and Citizen Participation and the Corps of Engineers.* Washington, D.C.: The Brookings Institute, 1979.

"McKeithen Asks $10 Million Aid." *New Orleans Times Picayune*, September 24, 1965.

McQuaid, John, and Mark Schleifstein. "Washing Away." *New Orleans Times Picayune*, June 23–27, 2002.

Memo for the Record. Subject: West Bank of the Mississippi River in the Vicinity of New Orleans, LA, September 17, 1990. RG 77, FY 00, N00021, Box 19, Westwego-Harvey File, Federal Records Center, Suitland, Maryland.

Meyer-Arendt, Klaus J. "The Grand Isle, Louisiana, Resort Cycle." *Annals of Tourism Research* 13 (1986): 463–64.

——. "Historical Coastal Environmental Changes: Human Response to Shoreline Erosion." In *The American Environment: Interpretations of Past Geographies*, edited by Lary M. Dilsaver and Craig E. Colten, 217–34. Lanham, Md.: Rowman and Littlefield, 1992.

Mississippi Valley Division Commander to Col. Michael Diffley (District Engineer). Memorandum, August 18, 1993. Project Management Files, Westwego to Harvey, folder 2, New Orleans District.

Monette, John W. "The Mississippi Floods." *Publication of the Mississippi Historical Society* 7 (1903): 427–76.

"Money Down the Drain." *New Orleans Times Picayune*, September 29, 2002.

Naomi, Al (Project Manager, U.S. Army Corps of Engineers). Personal interview, January 31, 2006.

National Oceanic and Atmospheric Administration, National Hurricane Center. "Preliminary Report: Hurricane Andrew, August 16–28, 1992." http://www.nhc.noaa.gov/1992andrew.html), viewed June 29, 2006.

——. "Preliminary Report, Hurricane Danny, July 16–26, 1997." http://www.nhc.noaa.gov/1997danny.html, viewed June 29, 2006.

——. "Preliminary Report, Hurricane Georges, September 15–October 1, 1998." http://www.nhc.noaa.gov/1998georges.html, viewed June 29, 2006.

———. "Tropical Cyclone Report: Hurricane Isidore, September 14–27, 2002." http://www.nhc.noaa,gov/2002isidore.html, viewed June 29, 2006.

———. "Tropical Cyclone Report: Hurricane Ivan, September 2–24, 2004." http://www.nhc.noaa.gov/2004ivan.shtml, viewed June 30, 2006.

———. "Tropical Cyclone Report: Hurricane Lili, September 21–October 4, 2002." http://www.nhc.noaa.gov/2002lili.shtml, viewed June 29, 2006.

———. "Tropical Cyclone Report: Tropical Storm Bill, June 29–July 2, 2003." http://www.nhc.noaa.gov/2003bill.shtml?text, viewed June 30, 2006.

———. "Tropical Cyclone Report: Tropical Storm Matthew, October 8–10, 2004." http://www.nhc.noaa.gov/2004matthew.shtml, viewed June 30, 2006.

National Park Service. *Jean Lafitte National Historical Park: Land Protection Plan*. New Orleans: U.S. Department of the Interior, National Park Service, 1984.

"Nearly Half a Million People Beat Betsy to Safe Area." *New Orleans Times Picayune*, September 10, 1965.

"New Orleans Sitting Duck for Major Hurricane. *New Orleans Times Picayune*, May 24, 1990.

"'No Growth' Line in Jeff Unresolved." *New Orleans Times Picayune*, April 30, 1980.

"Nothing's Easy for New Orleans Flood Control." *New York Times*, April 30, 2002.

O'Neil, Karen. *Rivers by Design*. Durham, N.C.: Duke University Press, 2006.

Orleans Levee Board to New Orleans District, Corps of Engineers. Correspondence, February 22, 1967. In U.S. Army Corps of Engineers, New Orleans District, *Lake Pontchartrain, Louisiana, and Vicinity, General Design Memorandum 2, Supplement 3, Chef Menteur Pass Complex*, Appendix B. New Orleans: U.S. Army Corps of Engineers, New Orleans District, 1969.

"Orleans May Escape Hurricane's Brunt." *New Orleans Times Picayune*, October 24, 1956.

Orsi, Jared. *Hazardous Metropolis: Flooding and Urban Ecology in Los Angeles*. Berkeley: University of California Press, 2004.

Owens, Jeffrey A. "Holding Back the Waters: Land Development and the Origins of the Levees on the Mississippi." Ph.D. diss., Louisiana State University, 1999.

Pabis, George. "Subduing Nature through Engineering: Caleb G. Forshey and the Levees-only Policy, 1951–1881." In *Transforming New Orleans and Its Environs: Centuries of Change*, edited by Craig E. Colten, 64–83. Pittsburgh: University of Pittsburgh Press, 2000.

"Parish Builds Levee in Marsh." *New Orleans Times Picayune*, September 17, 1998.

"Parish's Levee Debt Resurfaces." *New Orleans Times Picayune*, November 8, 1990.

Perez, Louis A., Jr. *Winds of Change: Hurricanes and the Transformation of Nineteenth Century Cuba*. Chapel Hill: University of North Carolina Press, 2001.

Piley, Anton, (Jefferson Parish Councilman) to Hale Boggs (Louisiana Congressman). Correspondence, April 17, 1968. National Archives, RG77, 72A, 3429, Box 1, Lake Pontchartrain file, College Park, Maryland.

Platt, Rutherford H. *Land Use and Society: Geography, Law and Public Policy*. Washington, D.C.: Island Press, 1996.

"President Promises Help." *New Orleans Times Picayune*, September 11, 1965.

"Protecting West Pank." *New Orleans Times Picayune*, July 5, 1980.

"Pumping Station to Be Completed Soon." *New Orleans Times Picayune*, October 14, 2004.

"Repairs Set for Buras Harbor Hit by Danny." *New Orleans Times Picayune*, September 27, 1997.

Reuss, Martin. "Andrew A. Humphreys and the Development of Hydraulic Engineering: Politics and Technology in the Army Corps of Engineers, 1850–1950." *Technology and Culture* 26 (1985): 1–33.

Sands, Thomas. (U.S. Army Corps of Engineers, retired.) Personal interview, January 1, 2006.

Saucier, Roger T. *Recent Geomorphic History of the Pontchartrain Basin*. Baton Rouge: Louisiana State University, Coastal Studies Series, no. 9, 1963.

Save our Wetlands v. Rush. Civ A. No. 75-3710 (ED LA Dec. 30, 1977).

Schroeder, R. H., Jr. (Chief Planning Division) to Commander Lower Mississippi Valley Division. Memorandum, January 16, 1992. Project Management Files, folder 2, New Orleans, District.

Seale, William B. (U.S. Army Corps of Engineers retired). Personal interview, March 14, 2006.

"Seawall Project off the Ground." *New Orleans Times Picayune*, July 7, 1994.

Secretary of the Army. *Annual Report Fiscal Year 1994 on Civil Works*, v. II. Washington, D.C.: Secretary of the Army, 1994.

———. *Annual Report of the Fiscal Year 1997 of the Secretary of the Army on Civil Works*, v. II. Washington, D.C.: Department of the Army, 1997.

———. *Letter from the Secretary of the Army: Lake Pontchartrain and Vicinity, Louisiana.* Washington, D.C.: U.S. Government Printing Office, 1965.

Seed, R. B., et al. "Investigation of the Performance of the New Orleans Hurricane Protection Systems in Hurricane Katrina on August 29, 2005," Draft Final Report, Report, Report No. UCB/CCRM-06/01, 2006. http://www.ce.berkeley.edu/~new_orleans/report/Chapter_1.pdf, viewed June 15, 2006.

Sewerage and Water Board of New Orleans. *Semi-Annual Reports*. New Orleans: Sewerage and Water Board, 1900–1940.

Shallat, Todd. "In the Wake of Hurricane Besty." In *Transforming New Orleans and Its Environs*, edited by Craig E. Colten, 121–38. Pittsburgh: University of Pittsburgh Press, 2000.

———. *Structures in the Stream: Water, Science, and the Rise of the U.S. Army Corps of Engineers*. Austin: University of Texas Press, 1994.

Sivori, Donald, (President Pontchartrain Shores Civic Association), to Hale Boggs (Louisiana Congressman). Correspondence, April 8, 1968. National Archives, RG 77, 72A, 3429, Box 1, Lake Pontchartrain file, College Park, Maryland.

"$6.7 Million Levee Bill Too Much for St. Bernard." *New Orleans Times Picayune*, March 17, 1990.

"Slidell Drainage, Flood-Control Work Hit Snags." *New Orleans Times Picayune*, July 30, 1999.

Solieau, Cecil (Retired Corps of Engineers, New Orleans.) Personal interview, February 22, 2006.

Spohrer, Gerald, (West Jefferson Levee District) to Terral Broussard (U.S. Army Corps of Engineers). Correspondence and attached Budgetary Procedures, August 17, 1992. Project Management, Civil Works Files, folder 1, New Orleans District.

Steinberg, Ted. *Acts of God: The Unnatural History of Natural Disaster in America*. New York: Oxford University Press, 2000.

Stine, Jeffrey K. *Mixing the Waters: Environment, Politics, and the Building of the Tennessee-Tombigee Waterway*. Akron: University of Akron Press, 1993.

"Study of Flood Problem Begins." *New Orleans Times Picayune*, October 26, 1956.

"Submerged Suburbs." *New Orleans Daily Picayune*, April 24, 1890.

"Tallying the Damage." *New Orleans Times Picayune*, September 29, 1998.

Team Louisiana, *The Failure of the New Orleans Levee System during Hurricane Katrina*. Baton Rouge: Louisiana Department of Transportation and Development, 2006.

Tobin, Graham, and Burrell Montz. *Natural Hazards: Explanation and Integration*. New York: Guilford Press, 1997.

Tommaseo et al. v. U.S. 80 Fed Cl 366. 2008.

Torres, Manuel, and Rob Nelson. "Unhappy Returns: Storm's Floodwaters Hamper Evacuees' Homecomings." *New Orleans Times Picayune*, October 5, 2002.

U.S. Army Corps of Engineers. *Annual Report of the Chief of Engineers on Civil Works Activities*. Washington, D.C.: U.S. Government Printing Office, 1992–2004.

———. *Budget Justifications*. Compilation of House Subcommittee Documents in digital format. Washington, D.C.: U.S. Army Corps of Engineers, Headquarters, 2005.

———. *Digest of Water Resources Policies and Authorities (EP 1165-2-1)*. Washington, D.C.: U.S. Army Corps of Engineers, 1999.

———. "Existing Pumping Station Stormproofing." January 2008. http://www.mvn.usace.army.mil/hps/existing_ps_stormproofing.htm.

———. "Hurricane Protection Improvements." June 2008. http://www.mvn.usace.army.mil/hps/hps_improve.htm.

———. "Key Feature of St. Charles Hurricane Protection Ready for Duty." News Release, January 8, 2001, RLINK http://www.mvn.usace.army.mil/pao/RELEASES/trepangnier.htm, http://www.mvn.usace.army.mil/pao/RELEASES/trepangnier.htm, viewed July 3, 2006

———. "Levees and Floodwalls." September 2006. http://www.mvn.usace.army.mil/hps/pdf/floodwall_levee_constr.pdf.

———. *Mississippi River Delta at and Below New Orleans, La.: Letter from Chief of Engineers, Army, Submitting Report on Interim Hurricane Survey of Mississippi River Delta at and Below New Orleans, La.* Washington, D.C.: U.S. Government Printing Office, 1962.

———. *Morgan City and Vicinity, La.: Letter from Chief of Engineers, Army, Submitting Report on Interim Hurricane Survey*. Washington, D.C.: U.S. Government Printing Office, 1965.

———. "MRGO Report Sent to Congress (News Release)." June 2008. http://www.mvn.usace.army.mil/hps/News%20Release_files/CPRA_ FINA_%20News_Release_2.pdf.

———. "New Orleans Projects—Actual Costs." Unpublished spreadsheet, 2005.

———. "Planned Permanent Pump Station Locations." June 2008. http://www.mvn.usace.army.mil/hps/hps_pumplocation_map.htm.

———. "Risk and Reliability Report." 2006. http://nolarisk.usace.army.mil/faqs.htm.

U.S. Army Corps of Engineers, Federal Emergency Management Agency, and the National Weather Service. *Southeast Louisiana Hurricane Preparedness Study*. New Orleans: U.S. Army Corps of Engineers, New Orleans District, 1994.

U.S. Army Corps of Engineers, Mobile District. *Disaster Operation, Mississippi Gulf Coast Following Hurricane Camille, 17–18 August 1965*. Mobile: U.S. Army Corps of Engineers, Mobile District, 1969.

U.S. Army Corps of Engineers, New Orleans District. *Abbreviated Transportation Model: Development and Format*. New Orleans: U.S. Army Corps of Engineers, New Orleans District, 2001.

———. "Caernarvon Freshwater Diversion Project: Fact Sheet," March 11, 1998, http://www.lacoast.gov/programs/Caernarvon/factsheet.htm, viewed July 5, 2006.

———. *Design Memorandum 1, Hydrology and Hydraulic Analysis, Part 1*. New Orleans: U.S. Army Corps of Engineers, New Orleans District, 1967.

———. *Design Memorandum 1, Part 1, Hydrology and Hydraulic Analysis—Chalmette*. New Orleans: U.S. Army Corps of Engineers, New Orleans District, 1966.

———. *Design Memorandum 1, Hydrology and Hydraulic Analysis, Part 3, Lakeshore*. New Orleans: U.S. Army Corps of Engineers, New Orleans District, 1969.

———. *Design Memorandum 2, East and West of Algiers Canal*. New Orleans: U.S. Army Corps of Engineers, New Orleans District, 1999.

———. *Design Memorandum 2, General Design, Citrus*. New Orleans: U.S. Army Corps of Engineers, New Orleans District, 1967.

———. *Design Memorandum 2, Lake Pontchartrain Barrier Plan, General Advance Supplement, Inner Harbor Navigation Canal West Levee, Florida Avenue to I.H.N.C. Lock*. New Orleans: U.S. Army Corps of Engineers, New Orleans District, 1967.

———. *Design Memorandum 2, Supplement 5b, New Orleans East Lakefront Levee, Paris Road to South Point*. New Orleans: U.S. Army Corps of Engineers, New Orleans District, 1972.

———. *Design Memorandum 3, General Design*. New Orleans: U.S. Army Corps of Engineers, New Orleans District, 1966.

———. *Design Memorandum 3, General Design, Supplement 1, Chalmette Extension*. New Orleans: U.S. Army Corps of Engineers, New Orleans District, 1968.

———. *Design Memorandum 3, Lake Pontchartrain and Vicinity, Louisiana, Chalmette Area Plan, General Design*. New Orleans: U.S. Army Corps of Engineers, New Orleans District, 1966.

———. *Documentation of Public Meeting, Lake Pontchartrain, Louisiana, and Vicinity Hurricane Protection Project*. New Orleans: U.S. Army Corps of Engineers, New Orleans District, 1984.

———. Fact Sheet (attached to memo of September 17, 1990). Federal Records Center, RG 77, FY 00, No. 0021, Box 19, Westwego-Harvey File, Suitland, Maryland.

———. Feasibility Review Conference, West Bank of the Mississippi River, Correspondence, October 27–30, 1993. Project Management Files, Westwego to Harvey, folder 2, New Orleans District.

———. *Final Environmental Impact Statement: Lake Pontchartrain, Louisiana, and Vicinity Hurricane Protection Project.* New Orleans: U.S. Army Corps of Engineers, New Orleans District, 1974.

———. *General Design Memorandum 1 (Reduced Scope).* New Orleans: U.S. Army Corps of Engineers, New Orleans District, 1989.

———. *Grand Isle and Vicinity, Louisiana: Final Environmental Impacts Statement.* New Orleans: U.S. Army Corps of Engineers, New Orleans District, 1974.

———. *Grand Isle and Vicinity, Louisiana: Draft, Phase I, General Design Memorandum.* New Orleans: U.S. Army Corps of Engineers, New Orleans District, 1978.

———. *Grand Isle and Vicinity, Louisiana: Phase I, General Design Memorandum: Beach Erosion and Hurricane Protection.* New Orleans: U.S. Army Corps of Engineers, New Orleans District, 1979.

———. "Grand Isle and Vicinity, Louisiana: Project Fact Sheet." http://www.mvn.usace.army.mil/pao/response/HURPROJ.asp?prj=lkpon1, viewed June 15, 2006.

———. *Grand Isle, Louisiana, and Vicinity (Larose to Vicinity of Golden Meadow), Design Memorandum 1, General Design.* New Orleans: U.S. Army Corps of Engineers, New Orleans District, 1972.

———. *History of Hurricane Occurrences along Coastal Louisiana: 1986–1997 Update.* New Orleans: U.S. Army Corps of Engineers, New Orleans District, 1997.

———. *Hurricane Betsy, 8–11 September, 1965: After-Action Report.* New Orleans: Army Corps of Engineers, New Orleans District, 1966.

———. *Hurricane Flossy, 23–24 September 1956, Memorandum Report.* New Orleans: U.S. Army Corps of Engineers, New Orleans District, 1957.

———. *Hurricane Hilda, 3–5 October 1964.* New Orleans: U.S. Army Corps of Engineers, New Orleans District, 1965.

———. *Hurricane Study: History of Hurricane Occurrences along Coastal Louisiana.* New Orleans: U.S. Army Corps of Engineers, New Orleans District, 1972.

———. *Integrated Final Report and Legislative Environmental Impact Statement.* New Orleans: U.S. Army Corps of Engineers, New Orleans District, 2008.

———. *Interim Survey Report, Lake Pontchartrain, Louisiana and Vicinity.* New Orleans: U.S. Army Corps of Engineers, New Orleans District, 1962.

———. *Jefferson and Orleans Parishes, Louisiana Urban Flood Control and Water Quality Management: Reconnaissance Study.* New Orleans: U.S. Army Corps of Engineers, New Orleans District, 1992.

———. *Lake Pontchartrain and Vicinity Hurricane Protection Project: Final Environmental Impact Statement*, New Orleans: U.S. Army Corps of Engineers, New Orleans District, 1974.

———. *Lake Pontchartrain and Vicinity, Lake Pontchartrain High Level Plan, Jefferson Parish Lakefront Levee, Design Memorandum 17.* New Orleans: U.S. Army Corps of Engineers, New Orleans District, 1987.

———. *Lake Pontchartrain and Vicinity, Lake Pontchartrain High Level Plan, St. Charles Parish North of Airline Highway, Design Memorandum 18.* New Orleans: U.S. Army Corps of Engineers, New Orleans District, 1987.

———. *Lake Pontchartrain and Vicinity, La.: Letter from Chief of Engineers, Army, Submitting Report on Review of Reports and Interim Hurricane Survey.* Washington, D.C.: U.S. Government Printing Office, 1965.

———. *Lake Pontchartrain, Louisiana, and Vicinity, General Design Memorandum 2, Supplement 3, Chef Menteur Pass Complex.* New Orleans: U.S. Army Corps of Engineers, New Orleans District, 1969.

———. *Lake Pontchartrain, Louisiana, and Vicinity, General Design Memorandum 2, Supplement 1, Rigolets Control Structure.* New Orleans: U.S. Army Corps of Engineers, New Orleans District, 1970.

———. *Lake Pontchartrain, Louisiana and Vicinity; General Design Memorandum 2, Supplement 6, St. Charles Parish Lakefront Levee.* New Orleans: U.S. Army Corps of Engineers, New Orleans District, 1969.

———. *Lake Pontchartrain, Louisiana and Vicinity, High Level Plan, Design Memorandum 22: Orleans Parish Lakefront Remaining Work.* New Orleans: U.S. Army Corps of Engineers, New Orleans District, 1993.

———. *Lake Pontchartrain, Louisiana and Vicinity Hurricane Protection Project, Reevaluation Study, Vol. 1, Main Report.* New Orleans: U.S. Army Corps of Engineers, New Orleans District, 1984.

———. *Larose to Golden Meadow, Louisiana Hurricane Protection Project: Mitigation Report.* New Orleans: U.S. Army Corps of Engineers, New Orleans District, 1987.

———. "Larose to Golden Meadow, Louisiana Hurricane Protection Project: Project Fact Sheet." http://www.mvn.usace.army.mil/pao/response/HURPROJ.asp?prj=lkpon1, viewed June 15, 2006.

———. *London Avenue Outfall Canal, Orleans Parish: Design Memorandum 19A, General Design*, v. I. New Orleans: U.S. Army Corps of Engineers, New Orleans District, 1989.

———. *Louisiana Coastal Protection and Restoration Technical Report.* New Orleans: U.S. Army Corps of Engineers, New Orleans District, 2008.

———. "Memorandum to Commander, Mississippi Valley Division, Annual Inspection of Completed Works, December 20, 2004, Region Wide Data, Annual Inspection of Completed Works." https://ipet.wes.army.mil/, viewed June 28, 2006.

———. *New Orleans Projects—Actual Costs Spreadsheet.* Unpublished. New Orleans District, 2005.

———. *New Orleans to Venice Fact Sheet.* August 24, 2004. http://www.mvn.usace.army.mil/pao/response/HURPROJ.asp?prj=lkpon1, viewed July 7, 2006.

———. *New Orleans to Venice, Louisiana, Design Memorandum 1, General Design Reach 1.* New Orleans: U.S. Army Corps of Engineers, New Orleans District, 1971.

———. *New Orleans to Venice, Louisiana, Design Memorandum 1, General Supplement 3, Reach C.* New Orleans: U.S. Army Corps of Engineers, New Orleans District, 1972.

———. *New Orleans to Venice, Louisiana, Design Memorandum 1, General Design Supplement 4, Reach B2*. New Orleans: U.S. Army Corps of Engineers, New Orleans District, 1972.

———. *New Orleans to Venice, Louisiana, Design Memorandum 1, Supplement 5 rev.* New Orleans: U.S. Army Corps of Engineers, New Orleans District, 1987.

———. *New Orleans to Venice, Louisiana, Hurricane Protection: Final Environmental Impact Statement.* New Orleans: U.S. Army Corps of Engineers, New Orleans District, 1973.

———. *New Orleans to Venice, Louisiana, Hurricane Protection Project: Final Supplemental Environmental Impact Statement, Supplement II, Barrier Features.* New Orleans: U.S. Army Corps of Engineers, New Orleans District, 1987.

———. "New Orleans Water Flows Through Davis Pond: Press Release," March 26, 2002, http://www.lacoast.gov/programs/DavisPond/index.htm, viewed July 5, 2006.

———. *1985 Hurricanes Juan, Danny, Elena*. New Orleans: U.S. Army Corps of Engineers, New Orleans District, 1987.

———. *Orleans Avenue Outfall Canal. Design Memorandum 19, General Design.* New Orleans: U.S. Army Corps of Engineers, New Orleans District, 1988.

———. *Orleans Parish-Jefferson Parish, 17th St. Outfall Canal, General Design, Design Memorandum 20*. New Orleans: U.S. Army Corps of Engineers, New Orleans District, 1990.

———. *Performance Evaluation of the New Orleans and Southeast Louisiana Hurricane Protection System: Draft Final Report of the Interagency Performance Task Force*, v. III: *The Hurricane Protection System.* New Orleans: U.S. Army Corps of Engineers, New Orleans District, 2006.

———. *Performance Evaluation of the New Orleans and Southeast Louisiana Hurricane Protection System: Draft Final Report of the Interagency Performance Task Force*, v. IV: *The Storm*. New Orleans: U.S. Army Corps of Engineers, New Orleans District, 2006.

———. *Performance Evaluation of the New Orleans and Southeast Louisiana Hurricane Protection System, Final Draft*. New Orleans: U.S. Army Corps of Engineers, 2006.

———. "Post Flood Report," in Southeast Louisiana Project, Jefferson, Orleans, and St. Tammany Parishes, Louisiana: Technical Report. New Orleans: U.S. Army Corps of Engineers, New Orleans District, 1996.

———. *Public Hearing.* March 13, 1956, National Archives, Record Group, 77, Accession year 1965, No. 285/98, Civil Works Planning Files, Mississippi River and Delta, Problems Caused by Hurricane File, College Park, Maryland

———. *Public Hearing on St. Bernard Parish, Louisiana*. December 15, 1965, National Archives, Record Group 77, Box 20, Survey Reports File, Fort Worth, Texas.

———. *Record of Public Meeting: Lake Pontchartrain, Louisiana, and Vicinity, Hurricane Protection Project, February 22, 1975*. New Orleans: U.S. Army Corps of Engineers, New Orleans District, 1975.

———. *Report on Hurricane Betsy, 8–11 September 1965*. New Orleans: U.S. Army Corps of Engineers, New Orleans District, 1965.

———. *Report on Hurricane Camille, 14–22 August 1969*. New Orleans: U.S. Army Corps of Engineers, New Orleans District, 1970.

———. *Review of Reports, St. Bernard Parish, Louisiana*. New Orleans: U.S. Army Corps of Engineers, New Orleans District, 1969.

———. *Southeast Louisiana Project, Jefferson, Orleans, and St. Tammany Parishes, Louisiana: Technical Report*. New Orleans: U.S. Army Corps of Engineers, New Orleans District, 1996.

———. *Southeast Louisiana Project: Jefferson Parish Technical Report*. New Orleans: U.S. Army Corps of Engineers, New Orleans District, 1996.

———. *Southeast Louisiana Project: Orleans Parish Technical Report*. New Orleans: U.S. Army Corps of Engineers, New Orleans District, 1996.

———. *Status Report on New Orleans District Projects*. New Orleans: U.S. Army Corps of Engineers, New Orleans District, 1983.

———. *Status Report on New Orleans District Projects*. New Orleans: U.S. Army Corps of Engineers, New Orleans District, 1990.

———. "Status Reports-Surveys." September 30, 1967, National Archives, Record Group 77, Accession No. 71A-2971, Folder 1507-01, MRGO, LA LVMVD, 1967, College Park, Maryland.

———. *Un-watering Plan: Greater New Orleans Metropolitan Area*. New Orleans: U.S. Army Corps of Engineers, New Orleans District, 2000.

———. "West Bank and Vicinity, New Orleans, Louisiana, Hurricane Protection Project, Project Fact Sheet." http://www.mvn.usace.army.mil/pao/response/HURPROJ.asp?prj=lkpon1, viewed June 15, 2006.

———. *West Bank of the Mississippi River in the Vicinity of New Orleans, Louisiana: Feasibility Report and Environmental Impact Statement (DRAFT)*. New Orleans: U.S. Army Corps of Engineers, New Orleans District, 1986.

———. "West Bay Sediment Diversion." http://www.mvn.usace.army.mil/prj/westbay/Fact.asp, viewed July 14, 2006.

U.S. Army Corps of Engineers, New Orleans District, to Commander, Mississippi Valley Division. "Annual Inspection of Completed Works." Memorandum, December 20, 2004. Region Wide Data, Annual Inspection of Completed Works, https://ipet.army.mil/, viewed June 28, 2006.

U.S. Army Corps of Engineers, Office of Chief of Engineers. *Annual Report FY 1983 of the Chief of Engineers on Civil Works Activities*. Washington, D.C.: U.S. Army Corps of Engineers, 1983.

U.S. Army Corps of Engineers, Waterways Experiment Station. *Effects on Lake Pontchartrain, Louisiana of Hurricane Surge Control Structures and Mississippi River-Gulf Outlet Channel: Hydraulic Model Investigation, Technical Report No. 2-636*. Vicksburg, Miss.: U.S. Army Engineer Waterways Experiment Station, 1963.

U.S. Congress, House. Committee on Public Works. *Hurricane Betsy Disaster of September 1965: Hearing before Special Subcommittee to Investigate Areas of Destruction of Hurricane Betsy*. 89th Cong., 1st Sess. September 25–26, 1966.

———. Committee on Public Works. *Southeast Hurricane Disaster (Hurricane Betsy): Hearing before Committee on Public Works*. 89th Cong., 1st Sess. October 13, 1965.

——. Committee on Public Works. *Southeast Hurricane Disaster Relief Act of 1965: Report from Committee on Public Works to accompany H.R. 11539*. 89th Cong., 1st Sess. October 13, 1965.

——. *Flood Control Act of 1965, House Doc. 231*. 89th Cong., 1st Session. 1965.

——. *Hurricane Betsy Disaster of September 1965: Hearings*. 89th Cong., 1st Sess. September 25–26, 1965.

——. *Lake Pontchartrain, Louisiana: Letter from the Secretary of War.* House Doc. 691, 79th Congress, 2nd Session. July 3, 1946.

U.S. Congress, Select Bipartisan Committee. *A Failure of Initiative: Final Report of the Select Bipartisan Committee to Investigate the Preparation for and Response to Hurricane Katrina*, 109th Cong., 2nd Sess., Report 109-377. Washington, D.C.: 2006

U.S. Congress, Senate. Committee on Appropriations. Subcommittee on Energy and Water Development. *Hurricane Preparedness: Hearing before a Subcommittee of the Committee on Appropriations. Special Hearing*. 103rd Cong., 2nd Sess. 1995.

——. Committee on Environment and Public Works. *A Bill to Authorize the Project for Hurricane and Storm Damage Reduction, Morganza, Louisiana to the Gulf of Mexico, Mississippi River and Tributaries: Report (To Accompany S. 2975)*, 107th Cong., 2nd Sess. November 4, 2002.

——. Committee on Homeland Security and Governmental Affairs. *Hurricane Katrina: A Nation Still Unprepared*. 109th Congress, 2d Sess., Report 109-322. Washington, D.C.: 2006.

——. Committee on Public Works. *Hurricane Protection Projects: Hearing before Committee on Public Works*. 85th Cong., 1st Sess., August 8, 1957.

——. Committee on Public Works and Transportation. Subcommittee on Water Resources. *Hurricane Protection Plan for Lake Pontchartrain and Vicinity: Hearing before Subcommittee on Water Resources of the Committee of Public Works and Transportation*. 95th Cong., 2nd Sess. January 5, 1978.

——. Lake Pontchartrain, Louisiana, Letter from the Secretary of the Army, Senate Doc 139, 81st Cong., 2nd Sess. February 22, 1950.

U.S. Department of the Army, Office of the Chief of Engineers. *Annual Report of the Chief of Engineers on Civil Works Activities, 1968*, v. II. Washington, D.C.: U.S. Army Corps of Engineers, 1968.

——. *Annual Report of the Chief of Engineers, 1969*, v. II. Washington, D.C.: U.S. Army Corps of Engineers, 1969.

——. *Annual Report of the Chief of Engineers on Civil Works Activities, 1970*, v. II. Washington, D.C.: U.S. Army Corps of Engineers, 1970.

——. *Annual Report of the Chief of Engineers on Civil Works Activities, 1973*, v. II. Washington, D.C.: Department of the Army, 1973.

——. *Annual Report of the Chief of Engineers on Civil Works Activities, 1974*, v. II. Washington, D.C.: Government Printing Office, 1974.

——. *Annual Report of the Chief of Engineers on Civil Works Activities, 1975*, v. II. Washington, D.C.: Government Printing Office, 1975.

———. *Annual Report of the Chief of Engineers on Civil Works Activities, 1976*, v. II. Washington, D.C.: Department of the Army, 1976.

———. *Annual Report of the Chief of Engineers on Civil Works Activities, 1982*, v. II. Washington, D.C.: Department of the Army, 1982.

———. *Annual Report of the Chief of Engineers on Civil Works Activities, 1985*, v. II. Washington, D.C.: Corps of Engineers, 1985.

———. *Annual Report of the Chief of Engineers on Civil Works Activities, 1989*, v. II. Washington, D.C.: Department of the Army, 1989.

———. *Annual Report of the Chief of Engineers on Civil Works Activities, 1992*, v. II. Washington, D.C.: Department of the Army, 1992.

———. *Annual Report of the Chief of Engineers on Civil Works Activities, 1994*, v. II. Washington, D.C.: Department of the Army, Corps of Engineers, 1994.

———. *Annual Report of the Chief of Engineers on Civil Works Activities, 1997*, v. II. Washington, D.C.: Department of the Army, 1997.

———. *Annual Report of the Chief of Engineers on Civil Works Activities*, v. II. Washington, D.C.: Department of the Army, Corps of Engineers, 2000.

———. *Annual Report of the Chief of Engineers on Civil Works Activities, 2004*, v. II. Washington, D.C.: Department of the Army, 2004.

U.S. Department of Commerce, National Oceanic and Atmospheric Administration. *Hurricane Andrew: South Florida and Louisiana, August 23–26, 1992*. Silver Spring, Md.: NOAA, National Weather Service, 1993.

———. *National Hurricane Operations Plan*. Washington, D.C.: U.S. Dept. of Commerce, National Oceanic and Atmospheric Administration [NOAA], 1982.

U.S. Department of Commerce, Weather Bureau. "A Model Hurricane Plan for A Coastal Community: National Hurricane Research Project, Report No. 28." Washington, D.C.: U.S. Department of Commerce, 1959.

U.S. Department of Defense, Chief of Engineers. *Annual Report, 1992, Civil Works Activities*, v. II. Washington, D.C.: U.S. War Department, 1992.

———. *Annual Report, 1995, Civil Works Activities*, v. II. Washington, D.C.: U.S. Department of Defense, 1995.

———. *Annual Report, 2000, Civil Works Activities*, v. II. Washington, D.C.: U.S. Department of Defense, 2000.

———. *Annual Report, 2004, Civil Works Activities*, v. II. Washington, D.C.: U.S. Department of Defense, 2004.

U.S. Department of Homeland Security, Office of Inspector General. *A Performance Review of FEMA Disaster Management Activities in Response to Hurricane Katrina*. Washington, D.C.: DHS, 2006.

U.S. Department of the Interior, Fish and Wildlife Service. *A Detailed Report on Hurricane Study Area No 1: Lake Pontchartrain and Vicinity, Louisiana*. Atlanta: U.S. Department of the Interior, Fish and Wildlife Service, 1962.

U.S. Executive Office of the President. *Federal Response to Hurricane Katrina: Lessons Learned*. Washington, D.C.: White House, 2006.

U.S. Federal Emergency Management Agency. *A Guide for Hurricane Preparedness Planning for State and Local Officials.* Washington D.C.: U.S. Federal Emergency Management Agency, 1986.

U.S. Fish and Wildlife Service to New Orleans District Engineer. Correspondence, May 15, 1968. In *U.S. Army Corps of Engineers, New Orleans District, Lake Pontchartrain, Louisiana, and Vicinity, General Design Memorandum 2, Supplement 3, Chef Menteur Pass Complex.* New Orleans: U.S. Army Corps of Engineers, New Orleans District, 1969.

———. Correspondence, November 29, 1969. In *U.S. Army Corps of Engineers, New Orleans District, New Orleans to Venice, Louisiana, Design Memorandum 1, General Design Reach.* New Orleans: U.S. Army Corps of Engineers, New Orleans District, 1971.

U.S. General Accounting Office. *Report to the Secretary of the Army: Improved Planning Needed by the Corps of Engineers to Resolve Environmental, Technical, and Financial Issues on the Lake Pontchartrain Hurricane Protection Project* (GAO/MASAD-82-39). Gaithersburg, Md.: U.S. General Accounting Office, 1982.

———. *Testimony: Army Corps of Engineers Lake Pontchartrain and Vicinity Hurricane Protection Project (GAO-05-1050T).* www.gao.gov/cgi-bin/getrpt?GAO-5-1050T, 2005.

U.S. War Department, Chief of Engineers. *Annual Report, 1992, Civil Works Activities*, v. II. Washington, D.C.: War Department, 1992.

———. *Annual Report, 1995, Civil Works Activities*, v. II. Washington, D.C.: War Department, 1995.

———. *Annual Report, 2000, Civil Works Activities*, v. II. Washington, D.C.: War Department, 2000.

———. *Annual Report, 2004, Civil Works Activities*, v. II. Washington, D.C.: War Department, 2004.

Van Heerden, Ivor, and Mike Bryan. *The Storm: What Went Wrong and Why during Hurricane Katrina.* New York: Viking Adult, 2006.

"Victory Declared on Levee Debt." *New Orleans Times Picayune*, August 2, 1990.

Vincent, J. Ross (President Ecology Center of Louisiana). Letter to the Editor. *New Orleans Times Picayune*, June 17, 1980.

"West Bank Levee Battle Restaged at Corps Hearing." *New Orleans Times Picayune*, April 18, 1984.

"West Bank Hurricane Levee Plan Criticized." *New Orleans Times Picayune*, February 28, 1980.

"West Bank Levee Starts amid Fanfare." *New Orleans Times Picayune*, March 23, 1991.

West Bank of the Mississippi River in the Vicinity of New Orleans, LA. Memorandum, September 17, 1990. Federal Records Center, RG 77, FY 00, No. 0021, Box 19, Westwego-Harvey File, Federal Records Center, Suitland, Maryland.

White, Gilbert. *Choice of Adjustment to Flood.* Chicago: University of Chicago, Department of Geography Research Paper, No. 93, 1964.

———. *Flood Hazard in the United States: A Research Assessment.* Boulder: University of Colorado, Institute of Behavioral Science, 1975.

———, et al. *Changes in Urban Occupance of Flood Plains in the United States.* Chicago: University of Chicago, Department of Geography Research Paper 57, 1958.

White, Richard. *Organic Machine*. New York: Hill and Wang, 1995.

Williams, S. Jeffres, Shea Penland, and Asbury Sallenger, eds. *Atlas of Shoreline Changes in Louisiana from 1853 to 1989*. Baton Rouge: Louisiana Geological Survey, 1992.

"Work Begins to Raise Lake Pontchartrain Levee." *New Orleans Times Picayune*, September 27, 2001.

Yenni, Michael J., (Jefferson Parish President) to Col. Michael Diffley (District Engineer). Correspondence, June 24, 1991. National Records Center, RG 77, FY00, No. 0021, Box 19, Westwego-LCA file, Federal Records Center, Suitland, Maryland.

Young, Tara. "Hurricane Levee Should Be Done by 2004." *New Orleans Times Picayune*, April 21, 2000.

Zebrowski, Ernest, and Judith A. Howard. *Category 5: The Story of Camille*. Ann Arbor: University of Michigan Press, 2005.

ARCHIVAL COLLECTIONS

NEW ORLEANS

Jefferson Parish. Police Jury Files. Gretna, Louisiana.

U.S. Army Corps of Engineers, New Orleans District. District Library.

U.S. Army Corps of Engineers, New Orleans District. Project and Engineering Files.

BATON ROUGE

LSU, Hill Memorial Library, Papers of Senator Russell B. Long.

FORT WORTH—NATIONAL ARCHIVES

Federal Court Records for Save Our Wetlands v. Early J. Rush, III, 75-3710 (E. La. Dist., 1977 and 1978.

Records Center, Record Group 77, Corps of Engineers.

Record Group 77, Corps of Engineers, N.O. District Project files.

WASHINGTON, D.C., AREA

U.S. Army Corps of Engineers, Office of History

National Archives: Record Group 77, Corps of Engineers.

Suitland Records Center: Record Group 77, Corps of Engineers.

INTERVIEW SUBJECTS

Carol Burdine
Terral Broussard
William Garrett
Thomas E. Harrington
E. R. Heiberg III
Victor Landry
William Landry
Al Naomi
Milton Rider
Thomas Sands
William B. Seale
Cecil Soileau

■ ■ ■ ■

INDEX